ROUTLEDGE LIBRARY EDITIONS:
POLLUTION, CLIMATE AND CHANGE

Volume 11

ENVIRONMENTAL
POLLUTION CONTROL

T0133947

ENVIRONMENTAL POLLUTION CONTROL

Technical, Economic and Legal Aspects

Edited by
ALLAN D. MCKNIGHT,
PAULINE K. MARSTRAND
AND T. CRAIG SINCLAIR

Routledge
Taylor & Francis Group

LONDON AND NEW YORK

First published in 1974 by George Allen & Unwin Ltd

This edition first published in 2020
by Routledge
2 Park Square, Milton Park, Abingdon, Oxon OX14 4RN

and by Routledge
52 Vanderbilt Avenue, New York, NY 10017

Routledge is an imprint of the Taylor & Francis Group, an informa business

British Library Cataloguing in Publication Data
A catalogue record for this book is available from the British Library

ISBN: 978-0-367-34494-8 (Set)
ISBN: 978-0-429-34741-2 (Set) (ebk)
ISBN: 978-0-367-36276-8 (Volume 11) (hbk)
ISBN: 978-0-367-36279-9 (Volume 11) (pbk)
ISBN: 978-0-429-34503-6 (Volume 11) (ebk)

Publisher's Note
The publisher has gone to great lengths to ensure the quality of this reprint but points out that some imperfections in the original copies may be apparent.

Disclaimer
The publisher has made every effort to trace copyright holders and would welcome correspondence from those they have been unable to trace.

Environmental Pollution Control

Technical, Economic and Legal Aspects

Edited by

ALLAN D. McKNIGHT
PAULINE K. MARSTRAND
T. CRAIG SINCLAIR

Foreword by LORD ASHBY

London. George Allen & Unwin Ltd
Ruskin House   Museum Street

First published in 1974

ISBN 0 04 309008 7 Hardback
ISBN 0 04 309009 5 Paperback

Printed in Great Britain
in 10 point Times New Roman type
by William Clowes & Sons, Limited
London, Beccles and Colchester

Foreword by Lord Ashby, FRS

Formerly Chairman of the Royal Commission on Environmental Pollution

The word 'ecology' has just celebrated its centennial. The first scientist to use it was Ernst Haeckel in 1873; there have for a long time been ecological societies, ecological journals, and professors of ecology; yet some recent writers about the environment give the impression that they have just discovered the word, and the subject. This is indicative of the tense, sometimes hysterical literature which has submerged environmental problems under a tidal wave of controversy. Any book which lifts environmental problems out of this turbulence and examines them clinically is to be welcomed; and this book (which does just that) deserves a special welcome because it puts the issues of environmental pollution into a fresh perspective.

There is a great deal we do not understand about the scientific aspects of pollution but we already understand quite enough to enable us to control pollution. Pollution control is hampered not by lack of knowledge but by lack of the will to carry it out. Politicians can do only what the public want and there is a limit to what the public will pay, or the inconvenience which they will tolerate, in order to enjoy clean air and water and land. So the control of pollution is primarily a political matter. Polluters will not, out of sheer altruism, refrain from putting their wastes into the environment; so they have to be persuaded by inducement or constrained by legislation; and both these acts need political decision.

Inducements cost money; legislation (if it is unpopular) may cost votes. And so the sound way to awaken public opinion about the control of pollution is not to publish sensational stories about incidents which have slipped through the machinery of control (poisoned fish or dumped cyanide or DDT in sea birds): it is to examine the machinery itself and the public attitude to the machinery. That is what many of the authors of these essays have done. There is an impressive amount of machinery and it has, on the whole, been effective. Britain is, by and large, less polluted today than at any time during the last 100 years. The machinery is clumsy and much of it is in need of an overhaul. But before overhauling it, it is necessary to know how it works. The value of this book is that it examines the administrative and legislative machinery and often skilfully diagnoses its deficiencies.

Pollution—as Bernard Levin once wrote—may be good for you. Indeed I cannot think of any pollutant, except radio-active fallout, which is not the by-product of some process which benefits society. The political problem is to balance the benefits against damage done by the by-product. It is impossible to do this numerically; therefore decisions have to be made on what politicians believe to be the value judgements of their constituents. The people of Manchester (for example) would doubtless like to have fish in the Mersey and a view of the Pennines. But what support will they give to orders and statutes or to higher charges on the rates in order that they may enjoy these environmental amenities? The answers to such questions as this one are encouraging. The British public are prepared to pay quite a lot for a cleaner Britain and to support quite tough legislation. But they need to pay more and to support legislation which is even tougher if the British environment is to be preserved for our grandchildren. The prospects of success for long-term conservation depend on a propensity among the public to put a higher value than they do at present upon the abatement of pollution. This is the sort of book which is likely to encourage that propensity because it explains, without jargon, how our environment can be preserved not by rhetoric, but by quiet logic; for in the long run logic prevails over rhetoric.

Contents

Introduction

The degradation of the human environment has been proceeding for centuries and has many causes. Only recently have a small group of scholars, notably Ehrlich and Commoner, vigorously stirred intellectual and public opinion to heed the earlier warnings of Carson [1], Rudd [2], and Moore [3] that man might be abusing his natural environment to a point of no return. The warnings of these scholars have been taken up by the mass media and in the early 1970s, environmental degradation has attracted the interest of many academics and created an increasing concern in the general public.

Environmental degradation lacks precise definitions of the environment itself and the causes of its degradation. There is agreement on certain fundamental factors such as pollution and wasteful use of natural resources. Urbanisation, due to increasing population and the declining use of labour in agricultural production, is the cause of providing many humans with an environment little better than what would be acceptable to certain animal and insect populations. At the United Nations Conference on the Human Environment in June 1972, much emphasis was placed on poverty, which exists in two-thirds of the world, as a prime cause of a degraded environment. Vigorous development policies were called for. This book has a strictly limited aim: to present a picture of pollution in the United Kingdom, against the background of the world community; its causes; the scientific and technological means of halting and then remedying it; the economic considerations relevant to the process of halting and reversing environmental degradation; the laws at present applicable and the need for their amendment; and (where appropriate) the administrative and judicial frameworks within which the laws operate.

Pollution has occurred for a long time, probably as long as man himself, but while his numbers were small in relation to a seemingly infinite world, effects were not noticeable. Early overt catastrophes produced by abuse of the environment are found in the deserts of the Middle East and India where deforestation, to provide more arable and grazing land, followed by over-cropping and over-grazing, produced waste land where formerly productive soil had existed. But the industrial revolution was the great accelerator in degrading the environment. Its direct effect was uncontrolled discharge of waste to the environment. Indirectly it led to a rape of natural resources and to a wide reorganisation of society with consequent urbanisation and squalid and close-packed housing. A second acceleration has taken place since 1945 due to increased population, the application of advanced technology without sufficient thought as to its possible side effects, and a vast increase in the advanced countries in production and consumption, producing an almost exclusive social objective of increased gross national product.

In Britain, pollution has been recognised as a problem for a long time. For example, the first legislation seeking to control air pollution in London dates from the thirteenth century. While this book was in preparation, the following letter was published by the *Brighton and Hove Herald* from its files of 1845.

125 YEARS AGO
(AUGUST 30, 1845)

"My Dear Mr Editor.—Is there any peculiarly good quality in the dirt and rubbish of Brighton, that they should be shot from the carts into the sea in the immediate vicinity of the most frequented bathing spots, so that the brine in which the bathers plunge is quite discoloured with a yellow dust? I am come here for a short time to afford my good wife (a sister of Mrs Caudle) the benefit, if benefit it be, of sea-bathing; my wife having been told that it would be very conducive to the health of her darlings, and render their skins beautifully clean, and give them brilliant and clear complexions; but on her return this morning, after the usual ablutions in the briny ocean, she found the skins of her fair daughters were covered with dusty particles, and required clean fresh water to purify them. She is somewhat doubtful whether bathing, under such circumstances, can be healthy, and she is quite positive it is most disagreeable.

"I am, my dear Mr Editor, yours truly,

LAVENDER LILLYWHITE"

One myth about environmental degradation should be dismissed. It is sometimes argued that environmental degradation is directly caused by a capitalistic organisation of society and that where Marxist socialist planning prevails the environment, being the subject of public ownership, will not be exploited since industrial undertakings are not motivated solely by profit motives. It should be pointed out that socialism alone is no guarantee of environmental protection: there has to be a determination to plan development and industrialisation in ways which will cause the least amount of damage, even if some of the 'benefits' have to be postponed until better technologies are available.

In preparing this book, we did not seek to define pollution for the line is difficult to draw, particularly on the boundary of misuse of natural resources. Instead we took five elements in the environment: namely air, inland water, land dereliction, the seas and noise. We invited authors to write about these subjects, concentrating on the scientific and technological, economic, legal and administrative factors and, where possible, the international considerations which are relevant. Marine pollution is of course largely a matter for international concern and hence action. Others will probably become so.

References

1. Rachel Carson, *Silent Spring* (Hamish Hamilton, 1964).
2. R. L. Rudd, *Pesticides and the Living Landscape* (Faber, 1965).
3. N. W. Moore, 'A synopsis of the pesticide problem', *Adv. Ecol. Research*, Vol. 4 (1967), p. 75.

Chapter 1

Technology and Pollution: A Case for Assessment

by CRAIG SINCLAIR

*Head of the Faculty in the Department of Innovation Management,
International Institute for Management of Technology, Milan*

Introduction

This introductory chapter attempts to discuss the background considerations applicable to the individual chapter discussions of particular pollutants and particular situations. In contrasting, as has been done in this book, the technological and the legal aspects of pollution, a balance is being demonstrated on an axis running from factual to arbitrary, between the impact of man on environment and his control of that impact. The impact arises from man's technology, that is the utilisation of scientific knowledge to control nature. This impact can be given some factual or statistical basis. The legal aspect introduces the notion of arbitration and is that of controlling man's nature itself.

That technology, in its adoption by industry for the increase of outputs, becomes a source of pollution seems to be a fact of industrial life. By the substitution of non-human energy sources for labour the load on the environment is increased. The increase in the scale of man's activities and the increase in his numbers pose seemingly major threats to the biosphere and even to human existence itself. The crucial question is whether this increased ability of man to manipulate his environment and to increase his production of material goods leads inevitably to a corresponding or greater increase in pollution levels [1]. For example, technological change can often bring with it an inadvertent improvement of a particular environmental ill. The substitution of nuclear power undoubtedly places less load on the air per unit of output. However, environmental policies must plan to make such amelioration purposely. This will require the involvement of disciplines other than those of technologists alone. In particular, unless the economic basis of pollution is demonstrated the acceptance of a purely environmentalist or conservationist argument will simply prevent any economic growth.

The debate about growth has been acrimonious, confused and remains unresolved. It has, for reasons to be proffered below, centred largely upon the recent experience of the United States, where impacts and reaction have been strongest. The concern is now world wide. Many shades of opinion and special interests are apparent, pessimistic ecologist [2] and optimistic technocrat [3], middle class preservers [4] of the status quo and Marxist defenders [5] of technological progress; each has debated with growth exponents from developing countries [6] and zero population growth zealots [7] from the industrialised West. For Paul Erhlich [8] the root of the trouble is the rapid growth in world population in the past few decades and the unavoidable increases over the next few decades. To Barry Commoner [9] the essential cause lies with technological practice itself and with the exploitation of technology for private gain. In these latter two the argument, as in almost all subsequent discussion, has been extended to include the entire Earth.

Taxonomies of world crises have been extensively elaborated [10, 11]. These have ranged from biological warfare to the destruction of civilisation by the motor car and the annihilation of entire ecological chains by chemicals and so on. The ultimate, and in a sense logical, outcome has been the attempt to build a mathematical 'model' of the world by Forrester and Meadows supported by the Club of Rome [12]. Among the ways in which the model predicted disaster for the world was a pollution crisis. In this model the level of pollution rose inexorably with increasing production and the persistent pollutants affected both crop yields, lowering food production, and acted directly on life spans to reduce these drastically. With existing knowledge such extrapolation, while having some admonitory effect, borders on the foolhardy. The model also implicitly denied the possibility of useful social control being introduced in time. The shortcomings both of the model and the data it used in the pollution sector (as well as other fields) have been pointed out at length by the team at the Science Policy Research Unit at the University of Sussex [13]. It is not the intention of this chapter, or this book for that matter, to review the arguments about potential global catastrophes in all their complexity, though it is hoped the volume will contribute to the debate. The aim is rather to describe, in particular detail, the mechanisms of the technologies involved and the legal procedures appropriate to their present and future control. The wider debate is mentioned here in order to make some linkage between the discussions of the individual chapters and to allow some comparative scaling of the problems.

The Scale of the Problem

The problems of pollution and environmental degradation are in a sense as large or as small as society chooses to make them. A neighbour's noisy car or the threatened destruction of a local beauty spot can be dealt with on a local *ad hoc* basis or as part of the larger, more universal, problem. Recently the idea of a general attack from technology upon the quality of everyday life [14] has had a widespread and almost spontaneous impact upon society. This has arisen because of man's comparatively powerful abilities (though patchily localised) to master and use nature and its forces for the satisfaction of his primary needs for food and shelter. Though achieved with strikingly varying degrees of success in different technological areas, this success has been sufficient in the past fifty to one hundred years to change the picture from a more or less delicately balanced and strenuously achieved utilisation of nature, to one where over-abundance and consumption distinguish certain large groups and societies today.

The concurrent advances of medical science have permitted a rapid population growth which presses hard upon available resources. The end, then, of a 'free fall' with regard to man's use of natural resources may have arrived.

Figures to illustrate the trends are not difficult to find. Man's increasing control of nature and use of its resources can be looked at first from the standpoint of his increasing use of extra-human energy and, second, of his increasing demand on the Earth for food. Each of these pollutes or degrades the biosphere to some extent, depending upon the rationale and techniques employed. Increasing demands can mean increased environmental disturbance. It will be convenient, since we have chosen to deal only with pollution, to dispose of the principal element of the latter first.

This facet of the problem is that of the population increase and its multiprong attack on land. The attack consists of a rising demand, in terms of actual numbers, on food resources and a concomitant pressure in the form of an increasing world-wide trend to urbanisation. Thus as industrialisation draws from the rural population, so increasing numbers put demands on agricultural production. For example, the urban population in the

USSR doubled between 1950 and 1969, rising from 69 to 134 million. In this way agriculture vies with industrial production and urban consumption as a producer of harmful refuse. Modern agriculture involves the use of pesticides which find their way into ecological chains with results which are wide ranging. Chemical fertilisers which may rapidly leach from the soil to watercourses add to the problem (see Chapter 5).

The statistics of the population increase are simple enough. There are about 3,700 million people in the world today and each year 70 million more are added. By the year 2000 the population will have almost doubled. This rate has been reached mainly by spectacular declines in mortality rates in the developing countries, some of which have growth rates as high as 3·5 per cent per annum against Western Europe's 1 per cent. This particular and basic cause of environmental disturbance is not dealt with in this volume. Not because the problem is not severe, but because the solution is simple—not simple in sociological terms or in political terms perhaps, but simple in statement. The answer to over-population is clearly the control of fertility. Thus discussion of pollution arising from growth would become nugatory if no population limits were reached. For example, the birth rate in England and Wales in 1870 if continued would have resulted in the population in 1970 being 140 million instead of 46 million. The discussion in this volume is carried out in the belief that this aspect of the problem has a separate existence and solution.

Turning now to the first theme, demand for increased energy supplies, which has been rising steadily at 3 to 5 per cent per annum, can be used as a source of illustration. Presently, 98 per cent of the work done in the world is provided by the combustion of fossil fuel and by hydro-electric power, with a small but growing contribution from nuclear power; while 100 years ago all but 5 per cent derived from muscle power [15]. If electricity power consumption is examined, the growth rate has always been about twice that for energy as a whole and runs at about 8 per cent per annum. The United Nations [16] predict that, given a world average population growth of 2 per cent per annum and a real per capita income growth of 3·5 per cent per annum, the demand rate will increase at the above value until 1980.

On these assumptions the global consumption of electrical energy may look like

$$
\begin{array}{lll}
1970 & 4{,}900 \times 10^9 \text{ kWh(e)} \\
1988 & 10{,}500 \times 10^9 & \text{''} \\
2000 & 33{,}600 \times 10^9 & \text{''}
\end{array}
$$

with total energy consumption rising at about half the electrical energy rate.

Such exponential rates of increase have two components relevant to consideration of the environmental effects. The first relates to the available resources and the effects of extracting them, and the second to the burden of residues and reject material involved in conversion. At present rates of increasing consumption the world's known supply of mineable coal will be consumed before 2100, though with decreasing consumption it may last until 2300 or 2400 [17]. A comparable value for the recoverable petroleum stocks is 2025. The world potential for water power is largely untapped and estimates are that it could be increased tenfold to 3×10^6 MWh(e). While there exist local possibilities for solar, geothermal and tidal power, the long-term prospect for power production must lie with nuclear power, hopefully fusion rather than fission, the potential energy of which is several orders of magnitude greater than the amount of chemical energy in fossil fuels. The growth of transportation will be a further contributor both to depletion of resources and to pollution. Chapter 3 considers the details of air pollution while the point to be made here is that the projected rates of increase of pollution from transport and other sources, in parallel with the growth figures above, demonstrates the extent to which the problem of pollution will grow if present programmes of control are not adaptable to future trends.

Definition of Impacts

An example from the Netherlands [18] is that of available water-cooling capacity for power production. Since the use of natural waters for this purpose imposes certain limits on allowable temperature rise (see Chapter 8) the total available to the Netherlands is roughly sufficient for 125,000 MW. By 2000 the demand will have reached 80,000 (at 7 per cent per annum increase) and pressure will be heavy on available sites near the water. Recourse will have to be made to cooling towers, bringing with them their own particular environmental problems—local effects on the micro-environment and possibly complaint about visual amenity. For countries with long sea coasts, for example, the UK, Italy and Denmark where few locations are further than one or two hundred miles from the coast, once-through sea cooling will provide the solution. For the Netherlands, however, suitable sites are few and the building of artificial islands, even when associated with other facilities like docks, for power station sites is uneconomic at present.

Three general points can now be developed using this example. First, that here the pollution discussed is 'thermal pollution'—the rejected heat from the process of producing electric power. This particular item is a latecomer to the usual list of pollutants, dust and chemicals in air or water, and noise. As such it illustrates the fact that items like rejected heat (or materials) achieve the title of pollutant only when they reach some threshold level at which they begin to appear socially unacceptable. The components of the list of pollutants are thus subject to change on the basis of social values as well as technical factors. Waste thermal heat has been a result of all electricity-producing processes for many years without complaint. This possibility of a succession of items to be considered as pollutants arises from the choices available both in social and technical action. Waste heat can be removed from the category by finding beneficial uses for it. Agricultural uses, fish farming, evaporation and concentration of sewage, and ice-free rivers are some of the suggested uses [19]. Few, if any, have proved economically feasible. The social acceptability of a pollutant level may depend upon its effect on morbidity or mortality levels or upon more subjective reactions. This confusion of the physical, measurable pollutant like SO_2, dust particles, DDT and even heat with subjective elements of nuisance like loss of beautiful views or crowded holiday resorts, causes the environment–growth debate to be confused. Technologies initially rejected, like railways or cast-iron construction, finally gain acceptance and even affection with public familiarity. No one could suppose that the physical pollutants—dirt and toxic chemicals—would ever become acceptable. The confusion of the two kinds of disamenity lingers in public attitudes, however.

The second generalisable point is that while the nature of pollutant problems often appears local it is becoming more and more true that abatement policies must very often be of an international character. Thus Germany, faced with having to find surface cooling capacity (having a small sea coast) for its power programme, will have a different rationale from the UK. The policies of Germany, Switzerland and France towards the use of the Rhine as a source of cooling must affect the Netherlands. Similarly the release of SO_2 from high stacks, which is the present policy in the UK, has been argued to cause pollution in Sweden.

The fact that the substitution of nuclear power for fossil-fuelled stations would at the present state of development increase the 'thermal' pollution problem (since the thermal efficiency is lower) allows the introduction of the third aspect to be discussed. This substitution would virtually eliminate the air pollution hazard of coal or oil combustion, even the radio-active component being higher from these fuels. Thus technology may be seen to enter here as a variable which can be altered to achieve improved environmental conditions. What has happened is that one type of pollution is substituted for another. On a wider level one type of risk to human life and environment has been replaced by another. Thus the choice might be seen as being between the risks. However, this is often difficult when these are different in nature and in time-scale and pose some awkward choices. Also the decisions as to technology are often made with regard first to strict economic results and only afterwards are the disamenity or welfare elements examined. This rise in apparent pollution problems and its coupling with changing preferences suggest that as the end of free fall [20] with regard to the exploitation of resources and environment is approached, some form of overall cost–benefit approach to technological change must be developed. The concept of a Technology Assessment function thus appears more and more attractive [21, 22]. The existence of very similar problems in the USSR and in the USA [23] suggests that the prevalent ideology or political cast matters less than the precise nature of the mechanisms directed towards the social control of technical innovation. The analogies which can be drawn between capitalist and socialist production with regard to pollution product must not be accepted, however, without at least reference to the superiority claimed [24], and partially disputed [25], for the Chinese experience.

The shape and function of any social control programme may thus depend in equal measure upon the historical background of the country adopting it and the special nature of technological risks. It will consequently involve complex and deep-seated social attitudes and responses as well as technical data. It will be useful to initiate some discussion of the Technology Assessment function by stepping aside to make a point about the springs of the US concern with environmental degradation.

Technology Assessment: National Approaches to Disamenities
The recent past has seen the enlargement of the 'standard of living' idea to encompass, besides consumption, items such as social security and leisure. The debate over environmental pollution, however, indicates that a still wider range of factors go

towards the social sense of well-being, increasing private affluence does not accord with increasing public effluents. This standard of living must include quality of living. Thus discussion of pollution and its control will inevitably be drawn into wider questions of the disadvantages of technical change. Clearly several approaches are possible—economic, ecological, technological—each sectioning the environment along different axes. The approach chosen in the following chapters is to describe in direct terms the legal and administrative institutions affecting pollution control in the UK and elsewhere, and the technical consideration and possibilities. These have been conveniently and conventionally separated in somewhat Aristotelean fashion as land, water, air and noise. In reality the division cannot be made so simply. Further examples can be given of the problem that technology allows us to convert a pollution in one medium, air (e.g. iron oxide in the flume from a steel furnace) to a pollution in another, water (when this is scrubbed out); or to turn a solid waste disposal problem to a gaseous one. Taken on the widest view this 'substitutability' of disamenities has a basic economic and technical rationale and has been mentioned above. Further examples might be local or specific—the removal of car noise, it is suggested, would result in greater pedestrian danger, the heightening of a power station chimney to lessen local air pollution levels may increase the danger to flights from the local airport. The concept implicit here is that cost, in some terms or other, attends all abatement and control techniques and the impossibility, in the vast majority of instances, of completely eliminating all disturbance from human activities. Thus with the benefit, the level of cost to be accepted becomes the issue.

The division referred to above, between legal and technological aspects, is also to some measure arbitrary and these 'cost' aspects must be considered against the socio-economic environment. This has implicit in it both social values and traditional economic wisdom. It is to some of these aspects that we now turn briefly.

In the United Kingdom, London's air is cleaner than it was ten years ago, and the Thames has fish in it once more. Why then, should we now have a national debate, relatively passionate, about pollution? How should we account for such passions within the area of public policy? Is the current international concern for environment related simply to the technological details, to sulphur dioxide levels in the air, the chemistry of detergents in the rivers, and the ecological chain that puts DDT into our fat, or is it symptomatic of some other, deeper, crisis in society?

J. K. Galbraith remarked recently that the crises which the

United States goes through at any time are just about five years ahead of similar crises in the other Western countries, and allows us just that five years in which to feel superior before we also are engulfed by the same problems. There are signs, too, that the developing countries are becoming concerned with the environmental results of development. In deeper analysis, however [26], it appears that the environment crisis is presently most acutely realised in the United States and may be a symptom there of a large foreboding. If this is so, is comparison between the United States and Europe and the rest of the world valid only on technical grounds, or is there a general basis for the similarity of response? Examining the US position a little more fully, an analysis of the content of a thousand or more textbooks used in the first eight years of American schooling during the nineteenth century, published a few years ago, reveals an attitude which demonstrates the roots of the dilemma. The almost universal contemporary attitude towards Nature revealed by the examination can be summed up as:

'Thus the nineteenth-century child was taught that nature is animated with man's purposes. God designed nature for man's physical needs and spiritual training. Scientific understanding of nature will reveal the greater glory of God, and the practical application of such knowledge should be encouraged as part of the use God meant man to make of nature. Besides serving the material needs of man, nature is a source of man's health, strength and virtue. He departs at his peril from a life close to nature. At a time when America was becoming increasingly industrial and urban, agrarian values which had been a natural growth in earlier America became articles of fervent faith in American nationalism. The American character had been formed in virtue because it developed in a rural environment, and it must remain the same despite vast environmental changes. The existence of a bounteous and fruitful frontier in America, with its promise not only of future prosperity but of continued virtue, offers proof that God has singled out the United States above other nations for His fostering care. The superiority of the American over older civilisations was shown by the endless frontier available. That Uncle Sam sooner or later will have to become a city dweller is not envisaged by these textbook writers, although their almost fanatical advocacy of rural values would seem to suggest an unconscious fear that this might be so.'

The United States has since become overwhelmingly urban and predominantly metropolitan and, in all probability, will move towards being overwhelmingly metropolitan. This has occurred since those textbook writers' days as a result of emigrant influx directly into the cities and by exodus from the land. The outcry against pollution may indeed be the half-strangled cry of a realisation that the Frontier image is lost, submerged in the trash of an urban-value-based technology, and if so it is nevertheless unhappily true that it is institutions developed in Frontier days that must cope with these technology-derived disamenities. Much of the trouble arises from the failure of the institutions to adapt quickly enough to the new technological, urban basis of production.

Such an analysis, even in the simplistic form given above, cannot apply to the present position with regard to environment in the United Kingdom or Europe. Further, in many areas, disamenity conditions with regard to public welfare are better than they were and public knowledge about many hazards is more extensive. The two principal areas of hazard to public health, i.e. from road death and from lung cancer, must be sufficiently visible both as to causes and levels. Both are areas of voluntary behaviour on the part of individuals and both are more or less accepted. These risk levels are much higher than those applying to the health risks derived from many pollutions. Public unease about air pollution and water pollution in the UK is not derived from a comparison of risk levels or from a sense of lost agrarian values, as in the US.

Part of the explanation for the diminished outcry lies, of course, in the longer history of the involvement of governments in Europe with problems arising from industrialisation [27, 28]. Control of land use and land-use change has a much longer history in Europe than in the US, though this has not been without its own particular costs [29]. This and the related body of legislation giving some measure of social control over the undesired effects of industrialisation and technical change has reduced the strength of upsurge of public concern with technology. A recent Japanese mission surveying the 'New Problems of Advanced Societies' around the world could find only the USA and possibly Sweden, as having a concern for environment similar to Japan's. In the latter the speed of growth has been so rapid that, as they admit, '[Japan] has experienced disastrous pollution directly affecting human lives' [30].

However, the very fact that public response in Europe, though widespread and fairly deeply felt, has been only partially channelled into effective pressure groups, could be argued to show the need for more adaptive response of government and official bodies to the threats posed by technology. Our slower growth and our slower reactions have made an acceptance of certain conditions which need not be accepted 'as the price you pay for growth'; more growth and higher quality of life is possible. For while, to take Japan as an example, Europe has not had its Itai-Itai or Minamata episodes, the longer time-scale of the introduction of innovation has left a residual legacy of technology-derived disamenity covering a wide area of human welfare. To limit the area only to pollution, it has been shown persuasively [31] that bronchitis levels are twice as high in urban as in rural areas in England and Wales. That piecemeal control of the social impacts through, in this case, the Clean Air Act has ameliorated the situation does not mean that other areas of disamenity do not still exist. It has been argued by the present author elsewhere that advanced technology in its industrial uses can be shown to carry a greater measure of social control than some older methods [32]. Further, it is often true [33] that the techniques exist to abate pollution but these are not used owing to insufficient mobilisation of public pressure or inadequate public education. On the other hand the Central Electricity Generating Board (who have a major potential impact in England), in evidence before the Select Committee on Nationalised Industries, maintained that it responded to public pressures on amenity questions by expenditures which matched the level of concern [34]. Where questions of visual amenity alone are at stake—land pollution—this may be the case. Here the damage is fairly apparent (see Chapter 7) and the cost of the remedies often easily determined. However, where the chain of events which link the polluter with the polluted is long and often hidden, as with many chemicals in our environment, then it is unlikely that public preferences can be so readily brought to bear. It seems clear that development of a national policy for the environment must proceed along several lines. It should be apparent from the examples in the foregoing and in the specialised chapters that while the physical 'causes' of pollution may lie in the increasing use of technological knowledge and artefacts in our society, the remedies lie in a complicated socio-technical web. Production technology pollutes but technology can abate the pollution. But at a cost; either of transferring it into another form which might be more or less acceptable, or as a determinable monetary prevention cost. This cost will increase as the desired standards are raised and with diminishing returns. The problem is to make the choices apparent and to determine appropriate levels from the expression of informed public opinion. The tasks of a method of Technology Assessment mentioned above might then be summarised as:

(a) The development of the physical methods of monitoring, measuring and determining the technical relationships involved in pollution situations.
(b) The development of methods of converting physical levels of the pollutant into a measure suitable for comparison with other disamenities (almost certainly this will mean in money terms).
(c) The development of mechanisms for the public display of these measures (costs) together with the benefits of proposed (and existing) technologies.
(d) The development of the institutions and organisations for the matching of public welfare preferences with the technological decisions.

It will be noticed that the above list of aims has moved from the purely technological to the socio-legislative and the remaining chapters of this book attempt to fill out the details of various pollutants in either aspect.

An Early Warning System

The first reports of the Royal Commission on Environmental Pollution [35] and of the US Council on Environmental Quality [36] make it clear that many of the more obvious hazards and disamenities are being dealt with; the levels are being reduced, economic measures of various types are being brought into operation to deal with pollution. The improvements achieved are to be praised and the attempts at public explanation of marginal cases encouraged. Much of the achievement in both cases comes from the piecemeal control of particular situations and the strength of existing laws and sanctions. In a fundamental sense, though, these two national approaches to the possible control of innovation differ radically. The second report of the Royal Commission says, apropos of the Impact of New Products, 'we want to promote immediate public discussion of . . . some early warning system . . . for the impact of new substances on the environment'. However, they believe that the assumption that 'every new product is guilty' until proven 'innocent' would 'inhibit desirable technological innovation' and might 'indeed be against the public interest'. Discussions have been initiated between the Confederation of British Industry and the Government on the subject of an early warning system on new forms of toxic wastes. The form of control is mentioned:

'We have in mind toxicological tests of the kind used already for the introduction of pharmaceutical products, pesticides, and food additives, together with other tests which, in the light of experience we already possess, might disclose dangers to the environment. This experience indicates, for example, that certain heavy metals, if they are likely to become combined with organic radicals—as in sludge on the sea bed—are a hazard to food chains; so are stable chemicals which are fat-soluble; so are stable chlorinated compounds and chelating agents (that is, substances which readily bring heavy metals into solution). While it would not be reasonable to regard substances with these properties as 'guilty until proven innocent' it is reasonable to regard them as 'under suspicion', which means that the industry which initiates their marketing should undertake sustained monitoring of their impact on the environment and should voluntarily publish the results of the monitoring.'

This arrangement is remarkably similar to that which applied to the production of drugs under the Committee on the Safety of Drugs. This voluntary vetting of pharmaceutical products has since been made compulsory under the provisions of the Medicines Act.

The proposal [37] to combine the five Industrial Safety Inspectorates and the Nuclear Installations Inspectorate is a further indication of a move towards a rationalisation of risk imposition and abatement.

On the other hand, the US legislature has shown a different general approach, no doubt because of the greater degree of public concern as discussed above. The passing of a Bill setting up an Office of Technology Assessment requires that certain innovations, covering a wider range of both technology and possible impacts, can be substituted to a full 'Technology Assessment'. The function of the Office [38] would be to:

1. identify existing or probable impacts of technology or technological programmes;
2. ascertain, where possible, cause-and-effect relationships;
3. identify alternative technological methods of implementing specific programmes;
4. identify alternative programmes for achieving requisite goals;

5. make estimates and comparisons of the impacts of alternative methods and programmes;
6. present findings of completed analyses to the appropriate legislative authorities;
7. identify areas where additional research or data collection is required to provide adequate support for the assessments and estimates described above; and
8. undertake such additional associated activities as the appropriate authorities specified below may direct.

The Board controlling the Office consists of twelve—six Senators, six Congressmen, evenly divided as to party—plus the non-voting Office Director. Assessments are initiated by the chairman of any House Committee or by the Board or by the Chairman in consultation with the Board. An Advisory Council composed of both public and technical members provides a link with non-government sources.

Clearly the operation of such an Office goes beyond the abatement of pollution and the task of proving the worth of its existence is a heavy one for the Office in its initial stages. Any similar capability which might be set up in European conditions would have to reflect the local situation. This is not the place to make the full analysis which would be necessary to weigh the need for such an authority in the UK.

The following chapters will perhaps impress on the readers' minds a firm preference one way or the other. That we must move towards some vetting process seems inevitable. If technological progress is to be harnessed for man's benefit then costs and benefits must both be demonstrated to an informed and participating public. The pitfalls are many and obvious—a strangulation of the inventive impulse, excessive caution leading to stagnation, a raising of public susceptibilities and fears. Yet if the idea of economic growth through applied technology has any basis in the scientific and rational attitude then it must be able to argue its case in an open and democratic way. The historical record is patchy but on the whole inspires confidence. The Micawber-like attitude of complacent optimism 'waiting for something to turn up' must not be replaced by a reaction towards innovation which, in the reverse way, waits for something to turn down.

References

1. W. Beckerman, 'The desirability of economic growth' in N. Kaldor (ed.), *Conflicts in Policy Objectives* (Blackwell, 1971).
2. B. Commoner, *The Closing Circle* (Cape, 1972).
3. J. Maddox, *The Doomsday Syndrome* (Macmillan, 1972).
4. R. G. Lee, *World Savers or Realm Savers* (Berkeley, School of Forestry and Conservation, University of California, 1970).
5. H. Rothman, *Murderous Providence: A Study of Pollution in Industrial Societies* (Hart-Davis, 1972).
6. J. A. Sabato and N. Rotana, 'Science and technology in the future development of Latin America', Paper to 'World Order Models Conference' (Italy, 1968).
7. H. E. Daly, 'Towards a stationary state economy', in Harte and Socolow (eds), *Patient Earth* (New York, Holt, Rinehart and Winston, 1972).
8. P. R. Erhlich, *The Population Bomb* (New York, Ballantine, 1968).
9. B. Commoner, op. cit.
10. J. Platt, 'What we must do', *Science*, Vol. 166, No. 3909 (29 November 1969), pp. 1115–21.
11. C. Quigley, 'Our ecological crisis', *Current History*, Vol. 59, No. 347 (1970), pp. 1–12.
12. D. Meadows *et al.*, *The Limits to Growth* (Washington, D.C., Potomac Assoc., 1971).
13. Science Policy Research Unit, *Malthus With a Computer* (Universe Books, University of Sussex Press, 1973 and Washington, 1973).
14. B. Commoner, *Science and Survival* (New York, Compass Books, 1966).
15. J. McHale, *The Ecological Context* (New York, Braziller, 1970).
16. UN annual statistics.
17. A. J. Surrey and A. J. Bromley, 'Energy', Chapter (8) in ref. 13 above.
18. K. J. Keller, 'Discharge of waste heat', Chapter IV in Kwee and Mullender (eds), *Growing Against Ourselves: the energy-environment tangle* (Lexington, Mass., Lexington Books, 1972).
19. Keller, op. cit.
20. C. Freeman, *Technology Assessment in its Social Context*, Studium Generale.
21. C. Sinclair, 'Technology assessment in the UK' in Cetron and Barlocha (eds), *Technology Assessment in Dynamic Environment* (New York, Gordon and Breach, 1972).
22. National Academy of Engineering, *A Study of Technology Assessment* (Washington, DC, USGPO, 1969).
23. M. Goldman, 'Environmental disruption in the Soviet Union', in *Man's Impact on Environment* (New York, McGraw-Hill, 1970).
24. Joseph Witney, 'Ecology and environmental control' in *China's Developmental Experience* (New York, Praeger, 1973).
25. R. A. Orleans and R. P. Suttmeier, 'The Mao ethic and environmental quality', *Science*, Vol. 170, No. 3963 (11 December 1970), p. 1173.
26. Fleming and Bailyn (eds) *Perspectives in American History* (Boston, Harvard University Press, 1973).

27. R. Gregory, *The Price of Amenity* (Macmillan, 1971).
28. Dept of the Environment, *Pollution: Nuisance or Nemesis?* (HMSO, 1972).
29. A. Briggs, 'A sense of place' in *The Fitness of Man's Environment* (New York, Harper Row, 1968).
30. Japanese Ministry of Foreign Affairs, Report on the Survey Mission discussed at the Symposium on 'New problems of advanced societies' (Tokyo, November, 1972) mimeo.
31. S. B. Lave and E. P. Seskin, 'Air pollution and human health', *Science*, Vol. 169 (1970), pp. 723-32.
32. C. Sinclair, *Innovation and Human Risk* (London, Centre for the Study of Industrial Innovation, 1972).
33. C. Sinclair, 'Environmentalism', Chapter 12 of Ref. 13.
34. Gregory, op. cit., p. 297.
35. Sir Eric Ashby (Chairman), 1st and 2nd Annual Reports of the Royal Commission on Environmental Pollution, HMSO, Cmnd 4585 (February 1971) and Cmnd 4894 (March 1972).
36. Environmental Protection Agency, Annual Reports, USGPO, Washington DC, USA.
37. Robens (Chairman), *Safety and Health at Work*, Cmnd 5034 (July 1972), HMSO.
38. Technology Assessment Act of 1972, 92nd Cong., 2nd session, report 92-1123, Calendar No. 1066, USGPO, 1972.

Chapter 2

Law and Administration

by ALLAN D. McKNIGHT
Lecturer, Civil Service College, London

'A law, in the proper sense of the term, is therefore a general rule of human action, taking cognisance only of external acts, enforced by a determinate authority, which authority is human, and, among human authorities, is that which is paramount in a political society. More briefly, a general rule of external human action enforced by a sovereign political authority.' [1]

'In one aspect law is essentially a restraint upon authority rather than the command of authority; its purpose and function is to limit rather than to reinforce political and economic power. In another aspect much of law is permissive rather than imperative; it does not consist of commands but of rules for securing desired legal consequences, conferring rights, creating obligations, and attaining other legal results. In this aspect law is a positive instrument rather than a negative restraint but an instrument of co-operation rather than of authority. The third element is that certainty in the law is alien to the function of law in a dynamic society; the life of the law is a constant interaction of factors of stability and factors of change.' [2]

Since earliest days man has been subject to many prohibitions and inhibitions in his individual behaviour. Perhaps in pre-history it is possible to contemplate man as an individual animal with unlimited freedom of behaviour. But at a very primitive stage the herding instinct led to the formation of tribes where the individual had to limit his freedom of behaviour in the interest of according maximum common freedom and enhancing the well-being of all members of the tribe. The relations of man to man in all the intercourse of life came to be regulated (before the evolution of any form of political authority) by a great body of maxims, arising partly under the influence of religion and partly out of the necessities of existence. In primitive society the remedy against an individual who breaks the tribal prohibitions and taboos is usually left as a matter of private revenge to the person who feels injured by the failure to observe tribal rules. The private revenge extracted is usually disproportionate to the original injury. As the tribe develops cohesively as a social entity, unlimited private revenge is superseded by some mode of regulated revenge; this is often encouraged by priests or other community leaders and becomes the norm because it has the support of the social conscience of the community. As Sir Henry Maine says, 'any expedients by which sudden plunder or slaughter was adjourned or prevented was an advantage even to a barbarous society' [3]. (But note that regulated self help still remains in some areas of modern English Law.)

Over centuries there were struggles in which one tribal leader tried to impose his will on an aggregation of neighbouring tribes. In the result there emerged the modern nation state, consisting of a numerous assemblage of human beings, generally occupying a fixed and certain territory, in which a sovereign power exercises authority over its citizens. (Indeed the word 'sovereignty' was first used by the early kings of France to reinforce their claims to primacy and authority over contending leaders.) The sovereign power in the state takes over the system of regulated revenge and tries, in most cases, to substitute therefor a system of granting redress to the person injured by the breach of a societal prohibition. This redress may be either restitution of the *status quo ante* or monetary compensation, the latter being far and away the more usual.

Not all tribal taboos are automatically justiciable in this way. At an early stage in the evolution of a state, the sovereign (acting either in person or by his appointed judges) consistently grants redress for the breaches of certain prohibited behaviour and refuses redress for others. Thus a distinction arises between legal obligations to which an individual must conform in his behaviour and moral obligations. Breach of the latter may be met by penalties varying from a contemptuous smile to complete social ostracism, but the courts are not interested.

The customary obligations which the courts will enforce constitute the common law of Britain; its source is custom and it is ancient; it evolved spontaneously in the popular mind as rules the existence and general acceptance of which were proved by their customary observance.

At the heart of any system of law is the concept of obligations (or duties) and their correlative rights. Thus if I sign a contract to sell my house, the purchaser is under an obligation to pay me the contract price and has a correlative right to receive a transfer of the title. I am under an obligation to transfer the title to him, with a correlative right to receive the purchase money. Such obligations and rights are known as obligations and rights *in personam*. If either party breaches his obligation, the other may seek redress from the courts, which, on proof of the breach, will grant redress, usually by the award of monetary compensation, but in exceptional cases, by ordering specific performance of the contract. Since few obligations and rights concerning pollution arise from contracts between individuals, little is said in this book about obligations *in personam*.

At the same time every person in a state is under innumerable obligations to forbear from performing certain acts or otherwise behaving in such a fashion as will or might cause injury to any other member of the community. For example, I must not assault others, nor defame them orally or in writing, nor murder them, nor rape them, nor drive my motor vehicle so negligently as to injure any other person or his property. These are known as obligations *in rem* and are owed to each member of the whole community. (The most important, in the context of pollution, is an obligation to forbear from any behaviour which might constitute a 'nuisance' to my neighbours.) Again, if there is a breach of any obligation *in rem*, any person affected by the breach may seek his remedy through the courts.

A legal obligation exists where one course of action is enforced, and the other prohibited, by organised society in the form of the state. The elements of an obligation are:

1. a person who is clothed with a right or who is benefited by its existence;
2. acts or forbearances which the person in whom the right resides is entitled to exact;
3. in many cases, an object over which the right is exercised; and
4. a person from whom those acts or forbearances can be exacted; in other words, a person under an obligation to act or forbear for the benefit of the subject of the right.

How far does custom and common law and its consequent system of rights and obligations encompass the problem of pollution?

First, mention should be made of attempts to control pollution by moral pressures. An example is the voluntary code of waste disposal practices, which recently was admitted to be a failure due to unscrupulous industrialists and fly-by-night entrants to the business of waste disposal. Despite the moral force of the code, according to repeated press reports, cyanide is still being dumped around Britain as irresponsibly as if it were old tin cans. This book does not attempt to discover how many codes of practices or 'gentleman's agreements' exist among small groups of citizens or between such groups and government departments, central or local. Rules for human conduct in relation to pollution can be assessed only by the laws governing such conduct, and equally, the extent to which they are enforced; in 1974, moral rules only weakly influence the behaviour of those engaged in polluting activities.

Much is said in the individual chapters of this book about the law of nuisance which is intimately connected with pollution. Its origins can be traced back to earliest times. Of all the common law obligations *in rem* it is the most relevant to pollution.

In origin, nuisance involved an interference with another person's use of his land. The interference could be with his natural rights, incidental to being the occupier of the land, or it could extend to his acquired rights, such as the right to fisheries in a river. A nuisance is created when deleterious things escape into another's land, for example, water, smoke, smell, fumes, gas, noise, heat, vibrations, electricity, disease germs, animals and vegetation. A nuisance must be unreasonable and there must be actual damage. It can become legal if unchallenged but public benefit is no defence.

On the whole, the law of nuisance (which was described as long ago as 1867 as being 'immersed in undefined uncertainty') [4] is full of so many intricacies as to be almost inapplicable to the circumstances of a modern industrial society in which a single householder (or a small group) is attempting to defend himself by legal action against the invasions of his use of his land or his enjoyment of his home by neighbouring industrial enterprises. Moreover, few private individuals or groups have at their disposal the means, financial and technical, to prove the necessary causative links between interference with their enjoyment of land and the operations of neighbouring factories, quite apart from the costs of bringing an action [5].

It should be noted that there can also be a public or common nuisance which is a criminal offence. It is an act or omission which materially affects the reasonable comfort and convenience of life of a class of Her Majesty's subjects. Action is taken by the Attorney General but there are few modern examples of such actions.

It is an interesting question whether the Government could have taken action on behalf of the two small communities in Wales adversely affected in their domestic lives by the noxious emissions from the plants of United Carbon and Rio Tinto Zinc Corporation [6]. Prosecution by the Attorney-General of the offending corporations would *prima facie* have been feasible since the reasonable comfort and convenience of life of a recognisable class of citizens were gravely affected. But the remedy has fallen into disuse because of a 'convention' that the Attorney-General will not act without first receiving an indemnity for costs. Perhaps some government, interested in controlling pollution, will ignore the convention, for the Attorney-General's suit to abate a public nuisance could be a powerful weapon in combating pollution.

Without a marked judicial leaning towards altering the common law to make it applicable to the circumstances of modern society, customary law tends to fossilise. Unfortunately this trend is more apparent in the United Kingdom than it is in the United States. In each country, it is necessary (but in varying degrees) for the legislative branches of government to change or add to the customary law.

This is the hallmark of an advanced legal system, that is, a legislative capacity to alter the law to meet changing circumstances. Whether its mode of exercise be democratic, autocratic or despotic, legislative enactments or decrees in the modern state are continually imposing new obligations (and to a lesser extent new rights) upon its citizens. Examples range from obligations to perform military service, to pay taxes, to send children to school, right through to an obligation to forbear from leaving a car for longer than two hours at one spot in a street. (Conversely the legislature often confers rights by way of social service entitlements but our interest is in the imposition by the legislature of obligations on all citizens to conform to prescribed standards of behaviour.)

These obligations attach to each individual in the whole community. They are *prima facie* obligations *in rem*. But the question arises as to where resides the correlative right? Does it reside only in the state or in each other individual citizen? Most of the obligations imposed by the legislature are in criminal form, that is, a breach of the obligation is punishable by a fine or imprisonment, although pecuniary penalties are numerically the most common. As we shall see later, the British courts have been ambivalent as to whether the correlative right is vested solely in the state or has been extended by the legislature to each member of the community injured by a person's failure to act in accordance with the obligation imposed on him by a statute.

Traditional common law crimes, like murder, rape, robbery and assault, originated in primitive times and severe penalties were meted out to those guilty of them. The development of the British Parliament in its modern legislative role roughly coincided with the onset of the industrial revolution. Parliament facilitated industrial activity by the technique of the joint-stock company, which allowed entrepreneurs to limit their risk by incorporating as an artificial person, to which the law gave legal personality and juristic status, and the ability of possessing rights and obligations. But as Parliament legislated to prevent the worst abuses of untramelled industrial expansion, such as the employment of children in coal mines, it faced the difficulty of imposing penalties on those guilty of breaches of legislative obligations, many of whom were incorporated companies. So the penalty almost invariably was a monetary fine.

Fines were fixed at modest levels. The industrial revolution commenced with no inhibitions whatsoever on the absolute freedom of the industrialist. He fixed the wages, determined the safety (or lack of it) of working conditions, arbitrarily employed people, including children, and often forced them to spend their wages exclusively in a shop owned by the factory. Labour was another commodity to be bought in the market place. Parliament acted slowly to curb such abuses partly because the overwhelming force in politics is a lethargic inertia against changing the *status quo* and partly because Parliament was exclusively composed of representatives of the rich agricultural and manufacturing classes.

When legislation was enacted, there was a danger that it would be ignored. It was necessary to appoint officials to ensure compliance with the law. Rarely has there been such an outcry in the House of Commons as occurred when legislation was introduced providing for the appointment of the first two factory inspectors to police obligations for industrial safety. The courts also played their part in maintaining the *status quo* so as to minimise the impact of socially protective measures on manufacturers. The most notorious example was the pernicious doctrine of common employment (enunciated in 1837) in which the judges rationalised that every contract of employment contained an implied term by which the employee absolved his employer from liability for any injury suffered as the result of the negligence of his fellow workers. The doctrine was only abolished by Parliament in 1948.

In the 1970s, the efficacy of our legal system in dealing with pollution must be judged by its capability to change or prescribe the law, to detect non-compliance and to punish non-compliance.

The keystone is Parliament which alone possesses the ability to impose obligations on citizens even though it may have delegated that power in a particular field to the executive. Parliament acts either by legislating itself—that is, by a private bill which is exceptional—or by approving a bill proposed by the executive. While many observers deplore Parliament's decreasing ability to influence the executive, it can by subtle means influence the Ministerial part of the executive to introduce legislation which, so far as pollution is concerned, amounts to imposing new standards on the whole mechanism of production, plus the not unimportant voting of funds to ensure the employment of sufficient staff and equipment to see that the new standards are observed.

The motivation of politicians (whether Parliamentarians or Ministers) flows from either (a) an inbuilt conviction of the need to act, (b) a conversion by private representations to the need to act or (c) a reaction to strong public feeling usually evidenced by, and often generated by, the press. On pollution, there is evidence from the press of grave public concern. What is not evident from the press is the level at which new standards should be fixed. Here the private representations which a Minister receives are all-important whether they be from his departmental advisers or from outside pressure groups of influential scientists. One suspects that the influential scientists rarely go further than presenting a need for action in principle, in the absence of sufficient prior research to suggest the new standards. This failure of the scientific community to produce precise recommendations makes it extremely difficult for politicians to act. At the same time, the political leader is subject to countervailing pressure from industry. Industrialists have on the whole opposed any regulation of their activities, whether it be English industrialists in the early nineteenth century opposing the appointment of the first two factory inspectors, or the Swiss pharmaceutical industry in the 1920s opposing the attempts of the League of Nations to control the traffic in narcotic drugs, or in 1970 the opposition of the nuclear power industry (often government-owned) to controls designed to ensure there is no diversion of nuclear materials from their generating plants to the manufacture of nuclear weapons. A pertinent example comes to mind in the attitude of British car manufacturers to the introduction of regulations to control the noxious content of exhaust emissions. As Professor Garner points out in Chapter 4, Parliament has already delegated authority to legislate on this subject to the executive although when it did so it had in mind more primitive considerations of safety of automobiles in crash circumstances. British car manufacturers have in their export models to observe American regulations. Yet the industry has made fervent representations as to its economic inability to provide similar models for the home market and, as a result, the regulations have been postponed for some years. It is certain that action through the Parliament or the Ministerial part of the executive will take some years to fructify.

But assuming Parliamentary legislation, either direct or through powers delegated to the executive, how effective is executive administration in observing compliance with the prescribed standards?

This is the second hallmark of a modern legal system, that is, there should be an executive administration which is efficient and effective in ensuring that there is compliance with the law, that violations are detected and that wrong-doers are then brought before the courts for the appropriate punishment. Each law requires a team of officials for this purpose, whether they be police, tax inspectors, customs officers, parking wardens or scientific inspectors. Such teams of officials need to be: sufficient in numbers; of high calibre; independent in mind and operation; prepared to prosecute wrong-doers; equipped with the latest scientific aids to maximise their efforts and minimise the number employed.

Responsibility for controlling the release of pollutants by industry and domestic users to the environment is divided between central and local government, each of whom employs inspectors. The calibre of the inspectors is no doubt as high as can be obtained at the salaries paid. It must be remembered that such inspectors deal with high-powered industrialists who are often extremely wealthy men, and the salary differentials are marked.

As to the size of the inspectorate, there must be some doubt. One recalls that the *Sunday Times*, 8 August 1971, reported a case where twenty-three police were employed to execute a search warrant directed to finding a purse containing 75p. Yet on the other hand, the Alkali Inspectorate, which is supposed to monitor the emissions from all industrial chimneys, has a total number of only twenty-six inspectors. River authorities and others responsible for attempting to control pollution have been similarly short staffed. As to independence of mind, the Alkali Inspectorate may again be quoted as an example. The inspectorate performs very little monitoring of its own and relies on plant records. It is shackled by its legislation which requires it to ensure that factories use the best *practical* means to minimise pollution. Few standards are laid down and the inspectorate works by persuasion. It is noteworthy that in the period 1920 to 1967 there were only three prosecutions. In 1970 two cases were heard involving two small concerns and the fines imposed were £50 and £75 respectively. As Jeremy Bugler concludes from these facts: 'the industrial super-corporations, who do most of the polluting in Britain, are not prosecuted' [7].

It must be recognised that no system of executive administration will provide an absolute guarantee against the emission of material to the environment. On the other hand, a rational and adequate effort, supported in full by the aids which science and technology have made available in recent decades, could lead to a very great improvement in the behaviour of the polluting factories. The first step is that all new industrial plants constructed must be approved from the point of view of incorporating the maximum anti-pollutant plant. Standards then need to be set and should not be capable of variation, as they were varied in the case of the Anglesey smelter of the Rio Tinto Zinc Corporation [8]. The next step in effective control is to have automatic monitoring instruments which are regularly checked by inspectors. Not only should there be adequate instrumentation to keep plant management informed, but this should be paralleled by tamper-proof instruments installed by the inspectorate, the results of which are read from time to time. Plant instruments need to be calibrated by the inspectors. Often automated sampling will be required and the inspector should remove a percentage of the samples (taken for plant management purposes) and have them independently analysed. Moreover, plant sampling should be checked for efficiency by feeding into the system standard samples prepared elsewhere. Nothing less than this, using statistical sampling methods to the greatest extent, will provide effective policing of any anti-pollution legislation.

The third test is the attitude of the judicial authorities. The courts, in their approach to what is euphemistically called 'white-collar crime' reflect the soft, timid approach of the executive administration [9]. While the executive administration is at all times unwilling to prosecute, when prosecutions are launched the courts are notoriously lenient. Within a few days in Brighton during the preparation of this book a defendant was sentenced to three months imprisonment for saying to a police inspector 'you Fascist pig'. In a second case the defendant, a medium-sized manufacturing outfit, had failed to provide the required safety machinery as a result of which a worker lost three fingers. In this case the factory was fined £100. Perhaps the attitude of the courts is epitomised in this small item from the *Guardian* of Wednesday, 5 July 1972.

'The Crown Court at Warwick yesterday halved fines imposed on a man for dumping drums of cyanide waste at children's play spaces. He was unsuccessful in his appeal against a prison sentence of 6 months for making false statements. The defendant was found guilty by Nuneaton Magistrates last month on two joint charges of dumping the cyanide. He was then fined £50 for each dumping offence; but sent to prison for stating that he was unemployed when he was in fact being paid for dumping the cyanide.
Judge Ervon Sunderland dismissed the appeal against sentence, but reduced the fines by half and gave him more time to pay.'

There is one respect in which the courts have been unduly restrictive and that is in their doctrine that breach by a person of a statutory duty or obligation does not always give a cause of action to any person injured as a result of the breach. The courts have adopted a rule that the intention of the legislature must be deduced by the courts from the statutory language creating the duty or obligation and no action will lie if the courts find a legislative intention that some other remedy, civil or criminal, shall be the only one available to an injured party. A method used by the courts in ascertaining the legislative intention is to ask whether the duty is owed primarily to the state or community at large and only incidentally to individuals. It would not seem that an action would lie at the suit of private individuals for damages arising from the release by a factory to the atmosphere of pollutant agents in excess of limits laid down by statute, although the fear of civil actions would be a more potent sanction in obtaining observance of the law than the normal sanction of a modest pecuniary penalty upon conviction.

What are society's values, as reflected by its courts? I personally find the judicial approach socially disgusting, ill-balanced and outrageous.

Many of the problems involved in controlling the pollution of the environment call for international action. This is self-evident in halting the pollution of the oceans. Moreover, certain pollutants are carried by an element of the biosphere across national boundaries so that pollutants released in one country cause damage in another. Thus Sweden [10], in relation to forest damage, and Norway, in relation to rivers and lakes, claim that damage has been caused by airborne pollutants originating in Britain, the Netherlands and West Germany. Any pollutants released to rivers in Western Europe will, via the North Sea, probably cause damage in littoral states. Moreover, the conditions of economic competition make it highly desirable that all industry, wherever situated, be subject to common rules and that all plant exported, whatever the country of origin or of destination, should incorporate common anti-pollutant devices.

The international legal system is at an early stage of development. There is no legislature but its absence is increasingly being filled by the use of multilateral treaties which fix standards and by ratifying a treaty, each state obliges itself to incorporate those standards in its domestic law. Similarly, despite the increasing number of intergovernmental organisations, the creation of international administrations to observe compliance with treaties is also at an early stage. In the field of judicial enforcement, again international society is at an early stage.

Overall the picture shows how much needs to be done in Britain in the legislative, administrative and judicial fields if governments are to cope with the massive problem of present day pollution. Most countries are in the same position. In the United States there has been much legislative activity, both state and federal, in the last few years but the new administrative agencies are still in the formative stage. By comparison Britain is not badly off for an institutional framework on which to build but this must not lead us to be complacent; the substantial job is still ahead. And national action needs to be accompanied by action to secure international agreements on a multilateral, almost universal, scale so that prudent regimes to reverse pollution are world wide. This will take time as witness the slow progress with the adoption of rules to preserve the oceans. There is room for neither complacency nor lack of urgency.

References

1. T. E. Holland, *Jurisprudence*, 13th edition (Oxford University Press, 1924). The first half of the chapter relies heavily on this book.

2. C. W. Jenks, *Law in the Modern Community* (Longmans, 1967).
3. On the subject generally, see any of the successive editions of Maine's *Ancient Law*.
4. *Brand* v. *Hammersmith Railway* (1867) LR 2 QB 223 at 247, quoted in J. G. Fleming, *The Law of Torts*, 4th edition (The Law Book Company of Australia, 1971). Fleming himself writes: 'Few words in the legal vocabulary are bedevilled with so much obscurity and confusion as "nuisance". Once tolerably precise and well-understood, the concept has eventually become so amorphous as well-nigh to defy rational exposition.' On the principles of tortious liability see Fleming, op. cit. and R. F. V. Heuston, *Salmond on Torts*, 15th edition (Sweet and Maxwell, 1969).
5. Even the revenue laws aid the polluter, whose costs of defending an action will be treated as a deductible business expense. Apart from a few exceptional cases, the plaintiff(s) will not possess this 40 per cent advantage.
6. The emissions were respectively carbon black and fluorines. For the facts and considerations in these two cases and other like cases, see Jeremy Bugler, *Polluting Britain* (Penguin, 1972).
7. Ibid.
8. Ibid.
9. On the subject of white-collar crime, see W. G. Carson, 'Some sociological aspects of strict liability and the enforcement of factory legislation', *Modern Law Review*, Vol. 33 (July 1970).
10. *Air pollution across national boundaries—the impact on the environment of sulphur in air and precipitation*, a publication by the Swedish Royal Ministries of Foreign Affairs and Agriculture (Stockholm, August 1971).

Chapter 3

Technical Aspects of Air Pollution

by R. S. SCORER

Professor of Theoretical Mechanics, Imperial College, London

The Essential Complexity

It is a commonplace that the weather varies around the year, from day to day, and often from hour to hour. Pollution emitted into the air moves with the air except for the larger particles of grit and dust, which fall out near to their point of origin. Inevitably, therefore, pollution exhibits a great variety of behaviour patterns.

We shall only be concerned with gases and smokes and mists of solid and liquid particles which are so small that their speed of fall through the air is negligible compared with the air motion. Pollution very soon becomes diluted by mixture with the air, so that although some gases (e.g. CO_2 and SO_2) are much heavier than air and others (e.g. water vapour, CO, NH_3) are lighter, they all move in the same way as the air into which they are mixed. Likewise smoke does not affect the motion of the air significantly on account of the weight of the particles, and the way smoke moves in the air is an excellent indication of how any pollutant, whether gas or particulate, would move.

Particles which are heavy enough to fall out on their own are also easy to remove from the air by centrifugal methods, and generally speaking there is no serious problem in removing them from flue gases before emission and we shall not be concerned further with them. Very small particles, on the other hand, are difficult to remove but are nevertheless the main cause of the pollution being visible, because their cross sectional area is greater for a given mass. Thus if as much as 95 per cent of the mass of particles carried in a stream of gas are removed before emission there is usually only a negligible reduction in their opacity. Electrostatic precipitators have then to be employed to remove the very small particles (below about 5 microns in diameter). If you look up at a polluted plume emerging from a chimney and particles fall in your eye, it is not those particles that you see but the micron-sized ones which travel with the gas.

The fact that gases such as CO_2 and H_2O have a different density from air only has a small effect on the density by comparison with the heat content. Hot gases rise with a speed proportional to the square root of the density deficit, and so the hotter they are the more rapidly they are dispersed upwards. At the same time the upward velocity is proportional to the square root of the size of the mass of hot gas, and so the effect is much greater when a large volume is emitted at one point than when the same volume is emitted at several places in small amounts. Thus the only case in which the buoyancy of effluent gases is important is when they are emitted much hotter than the air and in large amounts at one place, as at a power station.

In the absence of large buoyancy the position of the point of emission of pollution makes a great deal of difference. When emitted from a slim, tall, isolated stack, pollution is very well diluted before reaching the ground. When emitted from a short stack on a large building, or from a wide stack with low efflux velocity, the effluent is carried downwards in eddies behind the building (or the wide stack) and reaches the ground with much less dilution and much closer to the point of emission. Consequently the concentration of pollution observed at the ground depends enormously on the manner of emission.

It is important to emphasise, therefore, that the total amount of pollution emitted by a source is not a good guide to the effects it will have at ground level. Thus the SO_2 observed in streets is almost entirely due to the emissions from short chimneys on buildings in the street. A power station with a 150-metre stack situated in a town could well be emitting as much SO_2 as all the rest of the town put together, and yet the concentration it produces at ground level is quite undetectable by instruments at the ground because of the overwhelming concentrations produced by low level, smaller, cooler sources.

To the complexity of the weather we have to add the complexity of the sources emitting the pollution. Knowing that weather forecasting is a very crude art we cannot expect any accurate predictions of what pollution concentrations will be measured from a complex disposition of sources. Weather forecasting is only attempted on a daily basis with any sort of precision, and daily prediction of pollution is useless in planning. We therefore use the same kind of prediction as we do of the weather in the long term.

When we build a house we take note of the kind of extremes of weather it is intended to protect us from, and we accept that occasionally in a severe storm leaks or damage will occur.

In planning what emissions of pollution shall be emitted we can do no better than use experience as a guide to what is tolerable and try to avoid allowing new sources to be built which will produce objectionable results. This sounds so crude a method that it seems to imply that all sophisticated studies are of no practical use, but that is not a correct inference. We need to understand all the great variety of mechanisms which operate to produce all the effects we observe in order to know how to avoid the objectionable ones with the least expenditure. Only fundamental understanding can tell us how to design chimneys and how high it is worthwhile to build them. We need to avoid the occurrence of objectionable pollution incidents at night, in mid-winter, in summer anticyclones, in densely built-up areas and even in high winds whose gusts might bring down whiffs of pollution from a tall stack which otherwise causes no trouble.

Accurate predictions are impossible; predictions claiming any real precision should be suspected. No predictions can be better than extremely rough. The actual experience of any pollution source will, in normal weather, show variations of three orders of magnitude, and even on a day when short period whiffs of high concentration are observed the average for the day is often a tenth or even a hundredth of the high value.

Some sources emit more pollution in some kinds of weather. Most heating plants emit more pollution in cold weather although they will often emit more during a period of cool weather in a hot season than in the same kind of weather when it is average for the time of year; and they will consume more fuel in the evening than in the hours between midnight and sunrise at the same temperature. The kind of average that is appropriate to an industrial source whose emissions are independent of the weather is different from that appropriate to a heating plant.

In this essay no attempt is made to indicate how we deal with the problem of deciding what is an appropriate average, and how samples of observed pollution ought to be taken. It is merely emphasised here that total output is no reliable guide to the effect of a pollution source.

As if the complexity of the weather and the disposition and behaviour of the sources were not enough to make general principles difficult to establish and turn the art of pollution control into that of being a good local practitioner, we have to note that the effects of pollution are never simply in proportion to their concentration.

Thus CO (carbon monoxide) can kill, and indeed does if a car engine is run inside a closed garage; in small concentrations it has no known bad effect. Likewise SO_2 (sulphur dioxide) is poisonous, and has harmful effects on health in concentrations experienced close to a chimney; it is essential for the world's vegetation that there should be some of it in the atmosphere. CO_2 (carbon dioxide) is even more essential for plant life, and is needed all the time while SO_2 can be received adequately in intermittent doses (provided they are not so concentrated as to be damaging) and generally produces its beneficial effects by being oxidised and converted into ammonium sulphate $((NH_4)_2SO_4)$ which is removed from the air by rain and enters the plants through the roots.

The most objectionable components of car exhaust, the aldehydes, are only a minor pollutant and cannot be used as a measure of the concentration of car exhaust, while CO in exhaust in streets cannot be smelt and does no harm but is the easiest component to measure. All vehicles, however, do not emit the same proportions of all common car exhaust pollutants, and so it may lead to undesirable effects if legislation is tied to the most easily measured component because all efforts will then be directed to reducing it and not to reducing the most harmful ones.

Finally it must be noted that not only are many pollutants transformed while they are in the air into new chemical compounds, but may be absorbed into vegetation, on to stone and other surfaces, or removed by fog, rain or smoke particles. Most notable in this respect is SO_2 which has been widely used, not only as a tracer, because it is relatively easy to measure and is found in the flue gases of most large furnaces because coal and oil fuel mostly contain between $\frac{1}{2}$ and 4 per cent of sulphur, but is also used as a measure of the level of pollution because it is the only plentiful component of flue gases which is not relatively plentiful in the air far away from sources. CO_2 and H_2O are much more plentiful products of combustion, but there is so much of both in the air already that the plume from a chimney is soon lost if traced by them.

SO_2, on the other hand, is much less plentiful in the air in nature because it is so readily absorbed. Therefore it is a bad tracer because the concentrations observed will depend very much on the absorptive capacity of the surfaces over which the air containing the pollution has passed. Half the SO_2 in a stream of air might be removed from it as it passes through a dense hedge, when the concentration is fairly high, and the precise position of a sampling instrument can make an important difference to the readings obtained.

The Effects of Scale

In Figure 3.1 we may trace pollutants from their points of emission into the air and view them on a variety of scales.

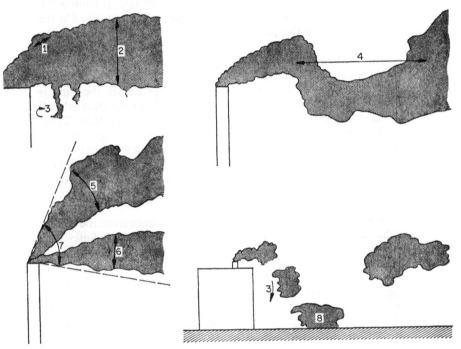

Figure 3.1   Patterns of behaviour of chimney effluent

Close to the chimney top we may see small eddies (1) which cause a plume made visible by smoke to widen. These eddies are small compared with the width of the plume (2) and may be caused either by the buoyancy of the plume which mixes with the air as it rises through it, or by eddies produced by the chimney (3) as they are produced on the downwind side of any obstacle not of stream-line shape.

Further away we notice that the plume becomes sinuous, showing that there are eddies (4) which push the plume first one way and then the other, and which seem to have a larger scale the further we move away from the chimney and the higher into the air the smoke is carried. These eddies are, of course, present close to the chimney, but the time-scale on which we think of the plume is smaller there. We think of the plume as existing for about the time taken for the smoke to travel from the chimney to the point we are considering. If we think of a point a few hundred yards away, during the time taken by the wind to travel that distance the plume at the chimney will sometimes be rising steeply from it (5), at others it will be carried almost horizontally away (6). On the larger time-scale the plume is therefore spread out more at the source (7).

If we are situated behind a building we may get occasional whiffs of almost undiluted pollution which are very objectionable when the chimney is of inadequate height (8); when the wind is gusty we may receive occasional whiffs from an isolated tall chimney which causes no trouble at all, most of the time. These whiffs may be unpleasant to people, and may do serious damage to vegetation, while building and decorative materials may suffer no harm from these whiffs because the reactions which affect them are very slow. On the other hand the materials may deteriorate in low concentrations maintained for a long time which have no effect on people and plants.

We soon see, therefore, that at a large distance from the source when the pollution is well mixed into the air it contributes to a background concentration which may do only small amounts of damage over a long time. The concentration at a large distance is scarcely at all affected by how high the source is in relation to buildings etc., but is very much affected by the size of a town, how many other sources there are near to it which contribute to the distant effect but only negligibly to pollution close to an individual source where concentrations may be very high.

The distant concentration is also affected by the absorption and removal rate, which is higher for SO_2, for example, than for odorous mercaptans or smoke.

The presence of smoke or other haze particles in the air has an effect on the temperature which is of no importance in a small volume of air but is of major importance if the air over the area of a large city is filled with smoke, and if the smoke remains for several hours. Not only does the particulate haze scatter sunshine back into space and reduce the amount of heating during the daytime, but it also radiates at night so that the air containing the particles is cooled directly. Normally, on a clear night, the ground radiates infra-red radiation into space and is cooled, and the air is cooled only indirectly by contact with the ground or by close proximity to it by radiation. The radiation emitted by the air is also absorbed by it so that the distance across which it can lose heat directly is only 2 metres or so.

The result is that when a dense particulate haze exists in the lower layers of air they tend to stagnate, in valleys particularly, and pollution emitted into the air therefore accumulates.

The Dilution of Pollution and its Removal from the Air

Some forms of pollution which do not fall out because they are gaseous or because the particles are too small are nevertheless deposited on solid surfaces. But this accounts for only a small fraction of the total, and usually pollution is removed first by being diluted into a much larger mass of air and then washed out by rain.

It is well known that the stronger the wind the more dilute is any air pollution, and this is the simple and direct result of the pollution being emitted into a larger volume of air than when the wind is light. For this reason all formulas for pollution concentration contain the factor (wind speed)$^{-1}$. Associated with the wind are eddies produced by the flow of the air over buildings, trees, and other ground roughnesses, and their intensity is proportional to the wind speed so that on their account the upward spreading of pollution is also increased with wind speed. This results in the concentration of pollution on any one occasion being proportional to x^{-2}, where x is the horizontal distance from the source. It also follows that the maximum concentration at the ground is proportional to h^{-2}, where h is the height of the source above the ground, and this result is independent of the ground roughness which determines the intensity of the eddies relative to the wind because that merely determines the distance at which the maximum ground level concentration will occur.

Many attempts have been made to represent the effect of buoyancy on plume dispersion by adding a quantity h^1 to the height h of the chimney. h^1 is called the thermal rise, and it is supposed that the plume will behave, as far as ground-level concentrations are concerned, as if it had been emitted without buoyancy from a chimney of height $h + h^1$. This turns out to be most unsatisfactory because h^1 depends on wind speed even in the most simple situations, and the only real use for formulas of this kind is for comparison of one chimney with another of different, but not too different, height.

The formulas fail because the eddies in the atmosphere are not simply determined by the wind speed and the ground roughness as they would be if the air were neutrally stratified, i.e. not stratified, but well stirred. In fact the air is either unstably or stably stratified most of the time, especially when the wind is not strong or fresh and the air has not been forcibly stirred by wind-produced eddies.

When the ground is colder than the air, as in the evening, the eddies are very much suppressed, and pollution emitted from low-level sources, such as houses, tends to be trapped in the lower layers, while that emitted from tall chimneys remains aloft and is not detected at ground level.

In the middle of the day, when sunshine warms the ground and causes strong convection currents in the air, the pollution from all sources is mixed rapidly into a very deep layer of air. This is the most important mechanism for the removal of pollution from air near the ground. When the convection currents are intense, especially when there are showers, there are downdrafts of cool cleaner air from above, and sometimes these bring down with them whiffs of pollution from high stacks.

Convection currents, and wind-produced eddies, do not carry the pollution above the level of the cloud base. At or very close to the cloud base there is usually a well-marked haze top, and to be carried higher the pollution must be inside an upcurrent entering a cloud. Only inside clouds are there any upcurrents penetrating above cloud base, and in anticyclones, when there is only very flat cloud, or even no cloud at all, the pollution is trapped below a stable layer (often called an inversion). To be cleaned of pollution the air must travel to a region where cloud is being formed, and in dry tropical regions such as California, haze may persist in the air for a month before this happens. For this reason there is a very serious air pollution problem when a region in these latitudes becomes highly industrialised.

In temperate latitudes it may only be a day and is seldom more than a week before air that passed over a source of pollution is carried above the cloud base. The pollution then acts preferentially as nuclei for the condensation of cloud droplets, although there are usually plenty of natural nuclei for the cloud-forming process to take place, and eventually when rain is formed, many droplets are aggregated into drops large enough to fall out, and the pollution is removed. Occasionally layers of pollution are seen at higher levels than the cloud base; these have been deposited there by clouds which have spread out at those levels and then evaporated again without raining.

The troposphere is the part of the atmosphere in which the weather occurs, and which extends up to the tropopause which is at about seventeen kilometres in equatorial regions but only at perhaps four or five kilometres in polar regions and is generally around ten or twelve kilometres in middle latitudes. In the troposphere therefore the air is frequently cleansed of pollution by rain, but in the stratosphere, above the tropopause, no rain occurs. Pollution is only removed from there when the air crosses the tropopause, which happens by a folding over of the tropopause so that some tropospheric air is deposited in the stratosphere and some stratospheric air is mixed into the troposphere.

If pollution is emitted into the lowest layers of the stratosphere it may be only a matter of a few months before most of it is washed out, but if it is emitted at much higher levels, above twenty-five kilometres say, it could be as much as ten years before most of it is removed, because the rate at which the air is overturned and eventually carried down to the tropopause is very slow at those altitudes. This is why debris from nuclear explosions, which is carried up to those parts of the stratosphere by the intense heat of the explosion, remains there for so long. Furthermore, most of it is rained out in temperate latitudes because it is there that most of the folding over of the tropopause takes place.

Pollution problems therefore arise when the air at the ground stagnates and is stably stratified. This happens particularly in valleys and at night and in dry regions of the earth, which are dry because the upper air in the troposphere is generally slowly subsiding and is producing no clouds and is warm enough to prevent convection currents from the ground from rising very high. In temperate latitudes serious pollution problems occur in the winter and in places enclosed by hills where there is a high population density with a large consumption of fuel.

Aircraft in the Stratosphere

The deposition in the stratosphere of debris from nuclear explosions is objectionable because the debris is itself objectionable. Aircraft exhaust on the other hand, like car exhaust, is a normal product of combustion of a kind which we can easily tolerate at ground level. Many cautionary suggestions have been made, however, that the introduction of the exhaust might have undesirable secondary effects. It is important not to exaggerate these because attention must never be diverted from serious real pollution problems on to others of no importance, especially when they are hypothetical. Unfortunately these suggestions have become mixed up emotionally with economic and other quite separate objections which some people have to the development of new high-flying aircraft, and the very fact that nothing has yet happened has been exploited. It is even argued that because we do not know the consequences we should not allow aircraft exhaust to be emitted into the stratosphere. When meteorologists assure the public that there is no danger and other scientists who are expert in other fields are not always convinced, the opponents of the aircraft argue that 'experts disagree'. Actually this situation is much simpler than is often imagined.

There are only two components of aircraft exhaust which are plentiful enough to have any possible effect; these are CO_2 and H_2O. Neither of these is of any importance in the troposphere because they are already plentiful there and aircraft consume only a small fraction ($1\frac{1}{2}$ per cent) of the fuel used in the world and are not a major contributor.

It has been argued that aircraft trails, which are clouds of water droplets condensed out of the exhaust and then turned into ice, might cause a shading of sunshine from the ground if they became too widespread. The trails do not persist in the stratosphere, and are only extensive and widespread over a very small fraction of those latitudes where they can persist for long enough to be a problem. For example, they sometimes cover an appreciable fraction of the sky in Colorado or Arizona where there is dense air traffic, but any possible effect on the weather is insignificant.

Since CO_2 is already plentiful in the stratosphere, the contribution from aircraft exhaust is of no consequence.

Water vapour exists in the stratosphere in much smaller quantities than in the troposphere. It has been estimated that if 500 SSTs each flew for seven hours every day and if the water vapour in their exhaust had an average residence time of ten years, the amount of water vapour in the stratosphere would be about doubled, after a few decades. If that happened the radiative balance would be altered, and the escape of heat from the troposphere would be reduced so that the air temperature at sea level would be increased by about half a degree Centigrade.

Before discussing the validity of this estimate we should set it in perspective. The temperature has fluctuated during recorded history by much more than this amount through causes not known for certain, but possibly connected with volcanic dust. In the last eighty years it appears that the total increase in CO_2 due to the world's fuel consumption has caused a rise in temperature of a degree or so, while in the last two decades the increase in smoke and dust has caused a reflection of sunshine which has produced a still greater decrease. This dust is mainly a tropical problem for there it persists longer in the air, and it is there that the production of dust has increased most due to the development of agriculture and exploitation of virgin land.

The estimate of half a degree due to moisture in the stratosphere from aircraft is certainly a gross overestimate anyway because most of the flying by SSTs is in the lower part of the stratosphere where the residence time of the exhaust emissions is only a year or less. The aircraft fly at a greater height as they use up more fuel and become lighter, so that on longer flights possible in the future a lesser proportion of the flying will be done at the greater heights. Even supposing the worst forebodings were true, the operation of the aircraft would be a simple thing to stop by international agreement, but it is certain that it will never produce a measurable effect because of the greater variations due to mechanisms already operating.

The other effect of exhaust in the stratosphere that has been suggested is that a reaction would take place which would reduce the amount of ozone (O_3) there. It has been suggested that if that happened, because O_3 absorbs ultra-violet (UV) light in the stratosphere, the amount of UV radiation reaching the ground would increase and that this would cause an increase in skin cancers. The UV radiation that is actually absorbed in the stratosphere, is however absorbed by ozone (O_3) at altitudes between about forty and sixty kilometres above the ground. At these altitudes no aircraft will fly. Even if they did, any depletion of ozone would quickly be made up by the absorption of solar radiation at those altitudes. The removal of all the ozone between twenty and ten kilometres would make a negligible difference to the amount of UV radiation reaching the ground, and the chemical changes postulated, if the amount of exhaust in the air were increased in the stratosphere, are not proved. They are merely suggested as possible and therefore to be guarded against because of our uncertainty; whereas there is probably no danger even if they were to occur.

The only serious insult to the environment caused by aircraft, other than noise which must be considered separately, is pollution near airports by aircraft on the ground. Aircraft engines do not burn fuel efficiently and cause many unpleasant odours and some smoke when operating at ground level where the pressure is greater and the speed less than at operational altitudes. The unpleasant consequences can only be avoided by a considerable improvement in airport operation so that aircraft are not kept waiting with engines running on the ground. There may be some moderate improvement in exhaust quality through technological advances to be made at the same time, but these cannot be guaranteed and will probably be slow in development.

The General Principles of Control

Most important in an industrial community is the control of atmospheres in which people work. It is relatively easy to monitor indoor air and to maintain certain required levels of freshness. The basis of this has always been to insist on the protection of work people and to ensure that processes are only operated when it is practicable to maintain the desired conditions.

This principle has been extended to the control of emissions into the atmosphere. At first it was necessary to control gross abuses of the environment. Now it is more a question of improving standards slowly in order that the air shall not deteriorate, and if possible that it shall be improved, as the population and consumption of fuel increases.

The 1956 Clean Air Act set out to replace sources of smoke with equipment which would perform the same task without making smoke. This was a practical approach because it was known that it was technically possible (and indeed fairly easy), and subsidies were made available to hasten the conversions. The administration was put in the hands of the local authorities, and the Government listed certain areas as being 'black areas', in which it was considered that the conversion was most urgent.

This is different from some legislation in the US which is applied on a statewide basis. For example the legislation on sulphur content in fuel applies to the whole of New York State. According to British practice this regulation would have been applied to the problem areas such as Manhattan first, because the method of tall chimneys is unworkable there.

For our present purposes it is important to emphasise that, while smoke which was known to be causing serious harm was reduced to only a small fraction, no attempt was made to reduce the emissions of SO_2 because no practical means was known of achieving this end short of reducing fuel consumption. The consumption of fuel was actually expected to increase, and with it the total emission of SO_2. The reduction in smoke emission was, however, expected to reduce the harmful effects of SO_2 and to increase the penetration of sunshine and reduce the stagnation of the air. The expectation has been realised and the dispersion of SO_2 has been greatly increased as a result.

At the same time all the larger sources of SO_2 have been subject to control of chimney heights. Since the highest concentrations have always been due to furnaces operating with inadequate chimney heights the highest concentrations have been avoided by proper chimney design.

Most of the increased fuel consumption has been in large furnaces, and particularly in power stations, and it is with these new installations that tall chimneys have been most notably developed. The newest power stations have chimneys around 200 metres high and even with powers of 2000 megawatts no harmful concentrations of SO_2 are produced at ground level.

At current world prices no commercially viable method of removing sulphur from fuel or from flue gases is known which would produce a useful sulphur product. Those methods of sulphur removal from oil which are practicable would produce an excess of sulphur which would have to be accumulated (i.e. dumped somewhere). Although it is probably desirable that extraction of sulphur from oil should be increased up to the level of the country's needs of sulphur this will not solve the problem caused by the presence of sulphur in widely used fuels. It seems certain that most of it will necessarily have to continue to be dispersed all over the world by emission from tall chimneys because any extraction would produce an undesirable accumulation of sulphur compounds. For example any wet washing of flue gases pollutes a body of water, and water pollution is usually more serious than air pollution.

It must be remembered that SO_2 is present naturally in the atmosphere. There is a cycle by which sulphur passes through all living beings and is emitted to the atmosphere from decaying vegetable and animal matter in the sea and on the land. It is a substance that is very quickly removed from the atmosphere and very readily digested in nature, and the only serious argument against its emission to the atmosphere is that an asset is being wasted. Unfortunately, in using oil fuel we cannot avoid extracting far more sulphur than could be used commercially.

It is quite contrary to the British philosophy of control to seek to prevent the emission of pollution primarily by prohibition and leave it to the courts to enforce the law. The objective is to avoid court cases as far as possible and only make emissions illegal where it is known to be, and has been made, practicable to avoid them. In the smoke control areas in towns the householders are equipped with grates which can burn authorised fuels smokelessly and the sale of unauthorised fuels is prohibited, and scarcely ever occurs. Consequently people are scarcely conscious of the existence of laws which might be broken.

The Criteria of Public Health

In the preceding section no mention was made of any attempt to establish standards of air pollution. The standards are set by the present levels and the practicability of improvement, and progress depends on public planning and efficient operation of local government.

If standards are set up which are specified in terms of the amount of pollution that is permissible in the air any use made of them implies an elaborate system of monitoring and of enforcement. The uncertainties in establishing an offence and of attributing it to a particular source are so great that in a country such as the USA, where the due processes of the law cannot be resisted, there is no end to the possibility of objection to a prosecution. Cases would be so long drawn out that enforcement would cease unless arbitrary formulas were adopted and imposed dictatorially and bureaucratically in order to frighten people into compliance with the law.

There does not therefore seem to be much practical use for specific standards. In any case they cannot be defined satisfactorily without either being far too restrictive because they seek to prevent all possible dangers, or because they are too lax and legalise certain obvious nuisances, or because they are so complicated as to be unintelligible to the general public.

The situation is simplified by the fact that public health is no longer a major reason for air pollution abatement. We know how to avoid most health dangers, and those who work in the public health field know that the worst menaces do not lie in the air pollution field. In particular the effects of cigarette smoking are so much worse that it is sometimes impossible to conclude from the figures available anything but that smoking is the major cause of ill health as far as breathing is concerned.

Car Exhaust

Although there are some components of car exhaust which are lethal in large concentrations, the exhaust is fairly innocuous in small concentrations. Our worry is not about the most lethal content, carbon monoxide (CO), because even in dense traffic no one could get as much CO in the blood as is achieved by the average smoker after a few cigarettes, and CO is probably not a major cause of the harm done by smoking.

Our concern should be directed at the unknown long-term effects of the less plentiful components of exhaust, namely the unburnt hydrocarbons and oxides of nitrogen, which might have long-term (twenty-year) harmful effects on health, and at the unpleasant components such as visible smoke and odorous aldehydes (and possibly mercaptans in diesel exhaust) and lead derived from 'anti-knock' compounds.

It is unlikely that exhaust concentrations will ever get higher than they are now, but the area over which the highest concentrations will be produced will increase as traffic reaches its maximum possible density on more roads. There is a real possibility of improving the quality of exhaust and making it less objectionable, but it is not obvious how the methods of doing this should be applied. It could be argued that all new cars should come up to a required standard, and if the standard can be achieved at a relatively low cost (say not more than 5 per cent of the cost of the car) it is reasonable to apply it universally. But if it is possible to achieve a much higher standard but at high cost (say 20 per cent of the cost of the car) it would be economically much sounder to require this standard to be reached in certain 'exhaust control areas'. In this way experience of the value of the regulations and of the effectiveness of the devices can be gained, devices of unknown ultimate value are not forced on to cars operating where there is no real problem, and the total effort is directed to where it is most needed. It would not be a serious problem to license 'clean exhaust' vehicles, and not permit others into the control areas.

Towering above these problems is the phenomenon of sunshine smog. This is the acrid haze that is produced in stagnant air containing pollution, mainly car exhaust, by the action of sunshine. It is known that both the unburnt hydrocarbons and the oxides of nitrogen play an important role. The result is considerable damage to vegetation, particularly to leafy crops such as spinach (whose production in the Los Angeles area has been made uneconomic) and to certain materials such as rubber, and an unpleasant smarting of the eyes with possible accompanying harm to other organs over a long period as yet unknown because the 'experiment' has not yet been carried on for long enough. Just as anyone can recover from one cigarette but get cancer after several years of smoking, so it must be suspected that unpleasant effects of sunshine smog might be long term.

Sunshine smog scarcely ever occurs in Britain because the air is not stagnant enough. When the weather is sunny the sea breezes replace the air too soon for a smog to develop. It is possible however that in one or two sheltered places, possibly in streets in cities, it might very rarely be detected for a short time in the middle of a summer afternoon. As in the case of most pollution problems, there are only two solutions: to stop the emission or to emit it from very tall chimneys into levels of the atmosphere where it is adequately dispersed and ultimately rained out. The second method is impossible in the case of cars; therefore the quality of exhaust must be improved or the internal combustion engine replaced.

It is important that emotional reactions to the problem should be controlled. The case for control in California does not justify similar controls in Canada or Britain, or even in Florida or Montana; the case for control in London does not justify it in Cornwall, or even in Kent. Nor must the advocates of clean air fall into the error of the anti-SST lobby and use erroneous arguments about pollution of the environment to support a case which is essentially economic or concerned with noise.

Odours and Special Chemicals

Odours present special problems. They are most objectionable if experienced intermittently, for if a person is exposed to a more or less constant concentration he becomes insensitive to them. After a few minutes of fresh air (whose freshness can be smelt) the odour will be noticed if it recurs. For this reason it is difficult for people working near the source, and possibly responsible for it, to remain aware of it and take action to stop it. Furthermore, as they become more dilute, odours actually change their smell, or to be more exact, the reaction of the human nose changes in quality as well as in intensity, so that it is not always easy to identify the source by smell.

But clearly many odours can be traced to their source. Even then it may turn out that the odour is caused by slight imperfection which has a negligible effect on the efficiency of the industry or the quality of the product and cannot be eliminated by the kind of controls that are employed for economic reasons. Nor are the odours intended, or planned. They are always liable to recur even when eliminated on a previous occasion, and must be recognised as a hazard in certain industries such as oil refining which ought to be sited accordingly.

Most odours are due to mercaptans, which are compounds of hydrocarbons with sulphur. They are easily destroyed by burning if they can be traced and removed, because SO_2 is the only perceptible product and that soon becomes imperceptible with dilution. Odours are difficult to detect, and still more difficult to measure. The only practicable method of measurement other than the nose is one which measures the sulphur content, and is therefore useless when there is a large amount of SO_2 about unless it is coming from the same source and can be taken into account accordingly. Most sources produce far more sulphur in the form of harmless SO_2 than as mercaptans, either from the same stack or from a nearby source.

There are other special chemicals such as fluorides which are produced particularly at aluminium and brick works, the latter of which also often produce unpleasant odours. Fluorides have the effect of weakening the teeth and unduly thickening the bones of cattle grazing on vegetation on which fluorides have been deposited and the only remedy is non-emission, or emission from very tall stacks.

Ammonia is often emitted from certain industries, and is about as harmless as SO_2 with which it combines after the oxidation of SO_2 to sulphuric acid to form an ammonium sulphate mist. This is a nuisance in areas subject to frequent natural fogs where chemical works are situated, for example on Teesside. The fogs are made more persistent and do not properly disperse but become a white haze when the wind is light. Ammonia is plentiful in nature, and is emitted from rotting vegetation and urine and animal excreta. It has been known to damage vegetation close to a hen battery which emitted it, but such high concentrations are rare.

Hydrogen sulphide is also produced widely in nature from decaying organic matter, particularly in the sea. It is harmless in naturally found concentrations and is quickly oxidised to SO_2. It is poisonous, and has been known to be lethal when breathed in high enough concentrations such as might occur in an industrial leak. Occasionally it has been dangerous in sewers.

There are some elements emitted in small quantities from the combustion of oil, about the effects of which little is known. Vanadium is the one most often mentioned.

The problem of the management of radio-active materials is outside the scope of this chapter except to say that those that are deliberately released in the atmosphere have a short radio-active life and cannot be a hazard.

Dust

A plume from a power station may be visible for one of several reasons. In very cold weather moisture may condense in it like in a natural cloud. At certain times it is necessary to blow the boiler pipes and accumulated ash is emitted in large quantities. Occasionally in oil-fired stations black smoke is emitted when the station is warming up. Some stations have a white plume of sulphuric acid mist: we need not be concerned about this because generally it represents only a small fraction of the sulphur anyway, it is visible because the particles are hygroscopic and grow in size. Most stations have too small a proportion of the SO_2 oxidised to SO_3 for this to be happening.

Many modern stations burn pulverised fuel, and even the most efficient electrostatic precipitators fail to remove the smallest ash particles which are hollow spheres of silica. There is also some small particulate ash from oil-fired stations. In both cases the ash is inert and harmless and is so widely dispersed as to be no nuisance.

Dust is emitted in much larger quantities at cement works. This is very difficult to prevent and is not due to carelessness (the dust being either raw material or cement itself from which the industry earns its keep). The precipitators are much more difficult to operate than in a power station because of the alkaline dust, the lower temperature and the higher moisture content in the gases.

Steel and other metallurgical works are probably the worst producers of dust. The source of red dust is the oxygen-lancing operation in which oxygen (or air in earlier installations) is bubbled through the molten iron to reduce the carbon content. This dust can be captured before emission, and indeed is so captured at Linz, in Austria, where stagnant air in the valley made its emission most obnoxious. But the damage to the land by steel works is often a much more serious environmental problem. At steel works blast furnaces occasionally emit black smoky dust, and mineral grinding sheds a brown dust. The region polluted is always the immediate neighbourhood. Improvements are as fast as local public opinion demands.

Air Pollution in Perspective

Although pollution of the air has been more widely publicised and the subject of more recent legislation than water pollution and the spoliation of the land surface, it is a problem neither as serious nor as difficult to deal with as the other two, if only because it disappears almost as soon as the source ceases to emit. Bodies of water may take decades and land centuries to recover from the careless activities of man.

We never attempt to process the air except in rather foolish situations such as when the air-conditioning plant of a building inhales the effluent from its chimney because of silly design. But we do have to operate water and sewage treatment plants, and we do have to restore the ground after mining etc.

The air cleans itself. It is also the most efficient disperser of unwanted material, and as such has been used by man since civilisation emerged with the control of fire. Incineration has long been the most effective method of tidying up: it distributes unwanted material in the form of gas and smoke over millions of square miles. Our civilisation is utterly dependent on the combustion of fuel, the use of atmospheric oxygen, and the transport away of the combustion products. We do the same in our own breathing and the biosphere is quite capable of digesting these products harmlessly. It is nonsense to say, as is often said, that we treat the air as a sewer. We do not do so because most of our effluents have been processed before emission because they are the products of combustion—the world's best 'purification' process. It is the exception for us to emit other products, and as far as we can we seek to disperse them, not confine them as in a sewer. It is clear that civilisation will have to continue to use the air as a vehicle for its products of combustion—and why not? It is the other products of the processes that are the cause of most trouble, and these are usually unintended.

The symbol of air pollution is smoke. We have come to recognise it as the greatest offence against clean air ever, for it was not completely combusted. We know how to prevent it, but there still remains a prevention job to be done in a few places.

Bibliography

W. E. Brittin, R. West and R. Williams, *Air and Water Pollution*, Proceedings of Summer Workshop (University of Colorado, 1970).

B. M. McCormac, *Introduction to the Scientific Study of Atmospheric Pollution* (Dortrecht, Holland, D. Reidel, 1971).

R. S. Scorer, *Air Pollution* (Pergamon, 1968).

———, *A Radical Approach to Pollution, Population and Resources* (Liberal Publications Dept, 1972).

———, *Air Pollution—Problems, Policies and Priorities* (Routledge & Kegan Paul, 1973).

———, 'Air pollution: its implications for industrial planning' in *Long Range Planning* (Pergamon, 1970), pp. 46–54.

———, 'New attitudes to air pollution—the technical basis of control', *Atmospheric Environment*, Vol. 5 (1971), pp. 1–32.

———, 'The interaction of local weather and air pollution problems', *International Journal of Environmental Studies*, Vol. 1 (1971), pp. 259–65.

A. C. Stern (ed.), *Air Pollution* (Academic Press, 1970).

Chapter 4

The Law Relating to Air Pollution

A. IN THE UNITED KINGDOM

by J. GARNER
Professor of Law, University of Nottingham

Introduction

Although a private landowner afflicted by smoke emissions from his neighbour's chimney or bonfire was (and remains) not entirely without remedy under the common law, as we shall see later, it is generally considered that the initiative in preventing or alleviating air pollution under modern conditions must be in the hands of a government agency or agencies, and in this respect Britain is no exception from other countries. Until after the 'Great Smog' in London of 1952, in which it is said that over 4,000 persons died within a period of two weeks, the powers given by Parliament to central and local authorities concerning air pollution were slight and virtually ineffective. As a consequence, however, of this dramatic event and the report of the 'Beaver Committee' [1], a comprehensive statute, the Clean Air Act, 1956, was passed. This statute, amended substantially by a section of the Housing Act, 1964, and the Clean Air Act, 1968, and supplemented by a mass of regulations, now provides a substantial piece of machinery for local authorities [2] to operate. The statutes and regulations are themselves explained in a large number of circulars and memoranda issued by the Ministry (now the Department for the Environment) [3] for the guidance of local authorities and members of the public. So far this, with the Alkali Act, constitutes almost the sum total of the law and practice of air pollution prevention; motor vehicle exhausts are controlled in an unsatisfactory manner by the regulations made under the Road Traffic Acts [4]. Although many local authorities have been active in implementing the provisions of the 'clean air' legislation, and the results are obvious in many of our cities, there are still serious gaps in the armoury of local and central authorities:

(a) there is virtually no control over sulphur dioxide emissions;
(b) emissions from chemical works are controlled by separate legislation, the Alkali, etc., Works Regulations Act, 1906, and the Alkali Inspectorate, on whom is placed responsibility for control over chemical and alkali works, is seriously under-staffed and seems to lack the initiative to prevent industrial forms of air pollution;
(c) the controls over emissions from motor vehicles are hopelessly inadequate.

We will return to these *lacunae* later [5]; we must first explain the scheme of the Acts of 1956 and 1968, designed to deal with 'smoke', grit, dust and fumes. But, however, it should be noted that the statute law with only minor variations, applies to Scotland as it applies to England and Wales. The 'common law' (para. 10) in Scotland is different.

Principal Provisions of the Clean Air Acts
These may be considered under a number of headings:

(a) the emission of dark smoke;
(b) controls over emissions of smoke, grit and dust from new industrial premises;
(c) the prevention of 'nuisances';
(d) the declaration of smoke control areas within which areas the emission of *smoke* from any chimney is totally prohibited.

We will now consider these provisions *seriatim*, but we must first explain what is meant by 'smoke' for the purposes of the Clean Air Acts. Subsequently we shall explain the provisions of legislation, other than the Clean Air Acts and of the common law, that is concerned with the subject of the prevention of air pollution.

'Smoke'
Smoke is explained by section 34(1) of the Clean Air Act, 1956, as including 'soot, ash, grit and gritty particles emitted in smoke', but this, clearly, does not provide a definition. Therefore until the superior courts have given us some assistance in this matter (and it is perhaps surprising that to date no case under either the 1956 or 1968 Acts has reached the High Court, either in London or in Edinburgh) [6], the word 'smoke' must be understood in its

ordinary dictionary meaning. Most dictionaries (in particular the Shorter Oxford) suggest that visibility is an essential characteristic of smoke, and it is therefore submitted that invisible vapours are not smoke, and that their emission is not capable of being controlled under the Clean Air Acts, expressly concerned as they are with smoke, grit, dust and fumes. Sulphur dioxide and carbon dioxide, for example, are not visible, and these common pollutants of the atmosphere in a highly industrialised country (sulphur dioxide in particular) are therefore not 'smoke'. The Clean Air Acts are therefore concerned principally with the carbonaceous products of combustion and more particularly of incomplete combustion, for with perfect combustion the whole of the principal constituents of fuel (carbon and hydrogen) are converted into carbon dioxide and water vapour.

'Grit' is not defined in the legislation, but is normally understood as meaning solid particles in the atmosphere in excess of 76 microns in diameter [7]; 'dust' is particles of a size from 1 to 76 microns in diameter [8], while the word 'fumes' means, according to the 1st Schedule of the 1968 Act, 'any airborne solid matter smaller than dust'.

'Dark Smoke'

'Dark smoke' is defined in section 34(2) of the 1956 Act as meaning [9] smoke which appears to be as dark as or darker than shade 2 on the Ringelmann Chart. This was a device produced by Professor Ringelmann in the last century, which consists of black cross-hatching on a white background of card, wood or other material, in such a manner that varying determined percentages of the white background are obscured. The chart is then held up by the observer and the general impression compared with the colour of the smoke emitted from the chimney under observation.

Section 1 of the 1956 Act makes it an offence [10] for the occupier of a building to allow or permit dark smoke to be emitted for periods in excess of those permitted under regulations made by the Minister [11] from any chimney of the building, subject also to certain defences provided for in the section (e.g. that emission was due to the starting up of a furnace from cold and all practicable steps had been taken to prevent emission of dark smoke). This section applies to emissions from chimneys serving industrial or trade premises, but not to dark smoke emitted otherwise than from a chimney at such premises (e.g. from the burning of used car tyres at a salvage depot). This gap in the section was rectified by *Section 1 of the 1968 Act,* which makes it an offence for the occupier of any industrial or trade premises to allow dark smoke from those premises (otherwise than by a chimney) to be emitted for periods in excess of those permitted under regulations made by the Minister [12]. These two sections must therefore be read together in practice; in the interests of simplicity, and so as to prevent technical legal arguments, it is somewhat unfortunate that the 1968 Act did not repeal the 1956 Act section and replace it with a comprehensive section providing for one offence covering all situations. It is also made an offence to emit dark smoke from the smoke stack of a railway locomotive engine [13], or from a vessel within waters not navigable by sea-going ships or waters within the seaward limits of the territorial waters of the United Kingdom within a port, harbour, river or estuary, etc. [14]. The exemption periods are not the same in cases of vessels, special regulations having been made to deal with them [15]. Ideally, it seems unfortunate that exemption periods have to be allowed at all, but in practice many furnaces cannot be operated without emitting some dark smoke for a limited period, and the exemption periods therefore represent a compromise.

Emissions from Furnaces

Section 3 of the 1956 Act and other sections of the two statutes deal with emissions of smoke, grit and dust in a different way, by attacking the source rather than (as in the case of dark smoke) attempting to control what is emitted from the top of the chimney. Thus, section 3 prohibits the installation in a building, boiler or industrial plant, of any furnace unless—so far as practicable—it is capable of being operated continuously without emitting smoke when burning an appropriate type of fuel. Section 3 of the 1968 Act requires grit and dust arrestment plant to be fitted to new furnaces (subject to exceptions specified in section 4), and section 6 of the 1956 Act makes it an offence to use a furnace installed after 1 June 1958, which will burn pulverised fuel, solid fuel or solid waste at a rate of one ton an hour or more, unless the furnace has been provided with grit and dust arrestment plant approved by the local authority.

The device of requiring local authority approval (which operates independently of other controls, such as town and country planning and the Building Regulations) applies also to section 3 of the 1956 Act, as a furnace which has been installed in accordance with approved plans is to be deemed to have satisfied that section. The height of a chimney serving a furnace constructed after the commencement of the 1968 Act also must be approved by the local authority [16], and it is expressly provided [17] that before they approve a particular chimney height the local authority must be satisfied that its height will be sufficient to prevent, so far as is practicable, the smoke, grit, dust, gases or fumes emitted from the chimney from becoming prejudicial to health or a nuisance.

There are also a number of 'machinery' provisions designed to assist in the monitoring of emissions from industrial premises. Thus, the Minister may make regulations [18] under section 4 of the 1956 Act requiring density meters to be installed at a furnace, and other regulations may [19] require recording apparatus to be installed for the measurement of emissions of grit and dust; these latter regulations enable a local authority to give directions to the effect that apparatus shall be installed at a particular building [20]. Further, under section 8 of the 1956 Act, a local authority may require information to be provided about particular furnaces and the fuel or waste burned in them. Also the Minister may by regulations prescribe limits on the rates of emission of grit and dust from the chimneys of furnaces in which solid, liquid or gaseous matter is burned [21], but no such regulations have yet been made.

These provisions together are reasonably comprehensive so far as the emissions of carbonaceous matter from industrial premises is concerned, but they are not concerned with other emissions. They also, in common with all the sections of the two Clean Air Acts, depend on vigorous implementation by local authorities and of course the making of appropriate regulations by the Ministry. Several sets of regulations have not yet been made and therefore the sections concerned are ineffectual, but the Ministry has certainly not been backward in giving advice to local authorities by way of circulars and memoranda [22].

Statutory Nuisances

Before the passing of the Clean Air Act, 1956, Parliament was content to leave the control of air pollution to the provisions of the Public Health Act, 1936, under which installations for the combustion of fuel in a manufacturing or trade process, which did not, so far as practicable, prevent the emission of smoke, and any chimney (*not* being the chimney of a private house) emitting smoke in such a quantity as to be a nuisance, were declared to be a statutory nuisance for the purposes of that Act [23].

These provisions were, however, repealed by the Clean Air Act, 1956, although section 31 of the Town Police Clauses Act, 1947, has so far been left unrepealed, which makes it an offence for any person wilfully to set or cause to be set on fire any chimney [24]. The 1956 and 1968 Acts clearly approach the problem of air pollution in a much more comprehensive manner than did the 1936 Act, but the idea of the statutory nuisance in relation to the emission of smoke has not been abandoned. Thus, section 16 of the 1956 Act, as amended by the 1968 Act [25], provides that smoke which is a nuisance to the inhabitants of the neighbourhood is to be deemed to be a statutory nuisance for the purposes of the Public Health Act, 1936. This requires a number of explanations:

1. *Exempted* from this provision is smoke emitted from the chimney of a private dwelling. Although the Beaver Committee estimated that over 50 per cent of all air pollution in this country was in 1954 caused by the emission of smoke from private dwellings, this particular form of pollution has been dealt with by the device of the smoke control area order (see page 69). Also exempted from the operation of section 16 is dark smoke emitted contrary to section 1 of the 1956 Act or section 1 of the 1968 Act; clearly because such emissions can be dealt with under those sections.

2. *Nuisance*. It must be shown that the smoke in question is a 'nuisance', a word which is presumably to be understood in the sense used by the common law, which is considered on page 74. Therefore where there is a common law nuisance by reason of the emission of 'smoke' [*not* chemical vapours] this may be dealt with in a private common law civil action, *or* under the present section.

3. *'Inhabitants'*. However, if the present section is used, it must be shown that the nuisance will be suffered by the 'inhabitants' of the neighbourhood. It is strange that the section does not use the word 'residents', and the distinction between inhabitants and residents is not clear, although it is submitted that inhabitants is a somewhat wider expression and would include persons working in the area as well as those there resident. 'Neighbourhood' also is a word incapable of precise definition or delimitation.

4. *Procedure*. On the first occurrence of a smoke nuisance within the meaning of the present section, the initiative will normally be taken by the local authority. They should thereupon serve [26] a notice [27] on the owner or occupier of the premises,

requiring him to abate the nuisance; if it is considered that the nuisance is caused by some structural defect in the premises, the notice must be served on the owner thereof [28]. If the notice is not complied with in the time specified for compliance in the notice (and this must be a reasonable time), the local authority should then cause a complaint to be made before the local magistrates [29]; when this is heard, the magistrates may impose a fine and also make a nuisance abatement order. Non-compliance with such an order constitutes a further offence, and also entitles the local authority to act themselves so as to secure an abatement of the nuisance, including a power to enter on the premises and carry out works if necessary, and they may then recover their reasonable expenses thereby incurred from the person in default [30].

If the local authority consider that a statutory smoke nuisance is likely to recur, they need not serve an abatement notice, but they may expedite this rather cumbrous procedure by making a formal complaint forthwith to the magistrates, asking for an abatement order, with the consequences above outlined [31].

Smoke Control Areas

Section 11 of the 1956 Act [32] empowers a local authority to make an order, which is then subject to confirmation by the Minister, declaring the whole or any part of their district to be a smoke control area. When such an order has come into force, it then becomes an offence if *any* smoke (not just 'dark' smoke) is emitted from a chimney of any building (including a dwelling-house) within the area, unless it can be established by way of defence that the emission of smoke in question was not caused by the use of any fuel other than an 'authorised fuel' [33].

The machinery of the smoke control area order really turns around the concept of 'authorised fuel', which means any fuel specified by the Ministerial regulations for the time being in force [34]. If an authorised fuel and only an authorised fuel is being used, no offence can be committed by the emission of smoke, and also particular types of fireplace may be exempted from the operation of the section [35]. Because of this emphasis on authorised fuels, it is also provided (in section 12) that works of adaptation to a fireplace, the alteration of a chimney, or the provision of gas or electric ignition for a fire [36], so as to enable an authorised fuel to be used and so avoid contraventions of section 11, are to rank for a grant from the local authority amounting to seven-tenths (or more at the discretion of the authority) of any expenditure thereby incurred by the owner or occupier of a private dwelling within a smoke control area [37]. A local authority may also serve [38] a notice [39] on the occupier or owner [40] of a private dwelling requiring him to carry out works of adaptation so as to avoid contraventions of section 11, subject to payment of a similar grant; and in default of compliance with such a notice, the authority would have powers to enter the dwelling and carry out the works specified in the notice, and then recover not more than three-tenths of their expenses thereby reasonably incurred [41].

Similar grants towards the cost of such adaptations may be made in respect of premises used for religious or charitable purposes (1936 Act, s. 14).

When an authority makes a smoke control order, notices have to be advertised in local newspapers and an opportunity has to be given for owners and occupiers in the area to make representations against the order to the Minister, who must then hold a local inquiry before an inspector appointed by him for the purpose [42]. Objectors must be given an opportunity of being heard at this inquiry and the Minister must consider the report of his inspector before deciding whether or not to confirm the order. Under the 1968 Act [43] the Minister may, after consultation with the local authority, direct a local authority to prepare proposals for the making and bringing into operation of smoke control area orders in their area, and if they default in compliance he may direct them to make such orders. These default powers have not yet been used, but at the end of 1968 there were no less than twenty-five local authorities in the 'black areas' [44] that had not then submitted any orders for the Minister's approval.

Nevertheless, the smoke control area provisions of the 1956 Act have been extensively used [45], and it is these provisions of the legislation that have probably had the greatest effect on the prevention of air pollution; the machinery, though complicated, has proved comparatively simple to operate and its administration has been eased because it takes the traditional public health form of order, inquiry and confirmation, followed by enforcement by means of persuasion [47] and the service of notices sweetened with generous grants [47] rather than by having to resort to prosecution at once. The difficulty in practice has been a spasmodic shortage of authorised fuels [48]; in some areas local authorities have been compelled to postpone or suspend the coming into operation of orders for this reason. On the other hand, the modern increase in the provision of electric, gas or oil-fired central heating has undoubtedly alleviated potential air pollution from domestic chimneys, whether or not smoke control areas have been declared.

Alkali Works

The legislation regulating emissions from chemical works (described in the legislation as 'alkali, etc. works') is quite separate from the Clean Air Acts hitherto considered, and is no concern of local authorities [49], being normally the responsibility of the Alkali Inspectorate, a centralised sub-department of the Department of the Environment.

The statute is still the Alkali, etc. Works Regulation Act, 1906, and this deals with four classes of works, namely:

(i) alkali works properly so called;
(ii) scheduled works; i.e. those listed in the first Schedule to the Act, as supplemented by the Alkali etc. Works Order, 1966 (SI 1966, No. 1143), and including a very wide range of chemical works;
(iii) cement works, in which aluminous deposits are treated for the purpose of making cement; and
(iv) smelting works, in which sulphide ores, including regulus, are calcined or smelted.

Any of these works must be registered with the Alkali Inspectorate [50], and certain standards laid down in the Act and enforced by the Inspectorate must be observed. Thus:

(a) every alkali works must be so carried on as to secure the condensation of 95 per cent of the muriatic acid gas evolved, and so that in each cubic foot of air or gas escaping into the atmosphere there is not more than one fifth of a grain of muriatic acid (hydrochloric acid gas);
(b) the owner of an alkali or scheduled works must use the best practicable means for preventing the escape of 'noxious or offensive' gases;
(c) any alkali waste must be so disposed of as not to create a nuisance;
(d) the owner of a scheduled works must also ensure that certain acid gases are condensed.

Some specified works may be excluded from the Alkali Inspectorate's jurisdiction and entrusted to the supervision of local authorities [51], but apart from these cases smoke emissions from alkali works (e.g. from central heating apparatus) are the concern of the local authorities and not the Alkali Inspectors [52]. Unfortunately there are not many (about twenty) alkali inspectors for the whole country and the statutory controls are not always as rigorously enforced as might be wished. Pollution of the atmosphere by chemical fumes and cement dust is therefore still pre-valent in many industrial areas, such as parts of Lancashire, Rotherham in Yorkshire and Scunthorpe in North Lincolnshire. On the other hand, conditions in areas where atmospheric pol-lution was primarily caused by smoke from domestic chimneys, such as London and the Home Counties, have very considerably improved since the passing of the 1956 Act.

Other Statutory Controls

These are at present only two: the control over the construction of new buildings or extensions to existing buildings under the Build-ing Regulations, 1972 [53] and the quite different controls over emissions from motor vehicle exhausts under the Road Traffic Acts and regulations made thereunder.

The Building Regulations. By regulation M2 it is provided that no appliance for heating or cooking is to be installed in a new building or extension to an existing building which discharges 'the products of combustion' into the atmosphere, *unless*:

    (a) the appliance is designed to burn as fuel either gas, coke or anthracite;

    (b) the installation complies with section 3 of the 1956 Act, in that so far as practicable it is capable of being operated con-tinuously without emitting smoke;

    (c) the appliance is of a kind for the time being exempt from the provisions of section 11 of the 1956 Act (which is con-cerned with the prohibition of the emission of smoke in a smoke-control area) by a Ministerial order made under section 11(4) [54]; *or*

    (d) the appliance has a bottom grate unsuitable for burning coke or anthracite but is designed so as to be capable of use with an alternative bottom grate which is suitable for burning such fuel.

Building Regulations are administered by the local authority, and before work is commenced on the construction of a building or an extension thereto, plans must be submitted to the authority and their approval obtained, whether or not planning permission has also been obtained. Approval must be refused if the plans are defective, or if they show that the proposed work would contra-vene any of the regulations [55]; moreover, if when the work is carried out there is a contravention of any of the regulations, proceedings may be taken in the magistrates' court for a penalty

[56], and the authority can require the work to be altered so far as may be necessary in order to comply with the regulations [57].

Road Traffic Regulations.   The Secretary of State is empowered by the Road Traffic Acts [58] to make regulations as to the construction and use of motor vehicles for the safety of road users; only incidentally is the secondary object of mitigating atmospheric pollution achieved, and dangers to public health caused by the emission of noxious vapours from vehicles is not really relevant to these regulations at all. The Construction and Use Regulations at present in force are those dated 1969 [59], and the regulations relevant to the present subject are as follows:

(a) *Reg. 24* provides that 'every motor vehicle shall be so constructed that no avoidable smoke or visible vapour is emitted therefrom'. This is of course not of great value so far as atmospheric pollution is concerned, as the regulation is concerned only with the construction of the vehicle, and in any event it is concerned only with visible emissions, whereas the most considerable pollutant from motor vehicle exhausts, carbon monoxide, is invisible.

(b) *Reg. 26* requires any vehicle using solid fuel to be provided with 'an efficient appliance for the purpose of preventing the emission of sparks or grit', and with 'a tray or shield to prevent ashes or cinders from falling on to the road'. This regulation, though clearly desirable, is not of great importance owing to the small number of vehicles so powered at present in use.

(c) *Reg. 84*—This regulation provides that a motor vehicle is not to be used on a road [60] if any 'smoke, visible vapour, grit, sparks, ashes, cinders or oily substance is emitted therefrom, if the emission causes or is likely to cause damage to any property or injury to any person' who is or may be expected to be on the road. This regulation is perhaps of greater value than regulation 24, but again it does not deal with carbon monoxide emissions, and it may be difficult in a prosecution to establish the actual or probable injury to other road users.

These regulations, apart from their own deficiencies, also suffer from difficulties of enforcement. In practice they depend for enforcement on the police forces and these overworked men are normally concerned with the more flagrant kind of breaches of the Construction and Use Regulations such as faulty lights, tyres or brakes, which may result in accidents, whereas the emission of fumes will rarely be a cause of accidents. Moreover, these regulations are concerned to control only visible vapours and not other noxious constituents of vehicle exhausts such as nitrogen oxides, hydrocarbons and lead compounds, as well as carbon monoxide. Obviously regulations against the emission of any form of noxious fumes from exhausts or the crankcases of motor vehicles should be aimed primarily at methods of construction, but a 'use' regulation requiring proper maintenance of injectors, etc., rigidly enforced, is also necessary.

Common Law

Apart altogether from the statutory controls on atmospheric pollution above described, the common law recognises the civil tort of 'nuisance', actionable at the suit of the person affected; and one of the causes of such a nuisance may be atmospheric pollution.

Grounds of the action. A nuisance is actionable by the plaintiff when the defendant does some act on his land which unduly interferes with the plaintiff's 'comfortable and convenient enjoyment' of his land [61]. No absolute standard can be applied, but it must amount to 'personal inconvenience and interference with one's enjoyment, one's quiet, one's personal freedom, anything that discomposes or injuriously affects the senses or the nerves' [62]. The acts complained of will of course vary according to the neighbourhood and the circumstances; but 'when it is said that a householder is entitled to have the air in his house untainted and unpolluted by the acts of his neighbour, by that is meant that he is entitled to have not necessarily air as fresh, free and pure as at the time of building the plaintiff's house the atmosphere then was, but air not rendered to an important degree less compatible, or at least not rendered incompatible, with the physical comfort of human existence' [63]. Thus, the creation of a stench [64], the causing of smoke or noxious fumes to pass over the plaintiff's property [65], or the raising of clouds of coal dust [66], have all been held to be actionable nuisances; there is no question of having to show that 'smoke' as defined by the 1936 Act was emitted, and it is not necessary to prove there was any financial loss or injury to health [67]. Sometimes a plaintiff may have to put up with a greater degree of discomfort by reason of the neighbourhood in which he has chosen to live, than would one living elsewhere; as was said by Thesiger, L.J., 'what would be a nuisance in Belgrave Square would not necessarily be so in Bermondsey' [68]. On the other hand, the emission of black smuts 'up to the size of a sixpence' was held to amount to an actionable nuisance, regardless of the nature of the locality [69].

Defences. The plaintiff can complain of a nuisance only if the acts of the defendant are such that they interfere with the reasonable use of his (the plaintiff's) land; 'a man cannot increase the liability of his neighbour by applying his own property to special uses, whether for business or for pleasure' [70]; it is, however, no defence for the defendant to show that the plaintiff 'came' to the nuisance. Ignorance of the existence of a nuisance may be a defence but not if the circumstances are such that the defendant ought to have been aware of the circumstances.

The remedy. The plaintiff may sue normally in the High Court, asking for damages and/or an injunction. If the nuisance in question is sufficiently serious and affects members of the public in the locality, proceedings may be taken for an injunction by the Attorney-General at the 'relation' and on the initiative of the local authority as for a public nuisance [71]; alternatively, where the local authority are of the opinion that proceedings for the abatement of a statutory nuisance under section 16 of the Clean Air Act, 1956 and sections 93–4 of the Public Health Act, 1936, would be inadequate, they may proceed in the High Court for an injunction in their own name (i.e. without the intervention of the Attorney-General) [72].

Conclusions

A remedy by way of common law proceedings may be most valuable to many a plaintiff, but the existence of a nuisance may not always be simple to establish, there is always a risk of heavy expenditure in civil litigation, and many prospective plaintiffs are not prepared to pursue their legal rights to the extent of taking court proceedings. Effective legal means of alleviating atmospheric pollution must therefore, it is submitted, continue to depend on the initiative of central and local authorities. Much has been achieved since the 'Great Smog' of 1952, especially by the device of smoke control area orders, but much remains to be done, and improvements in the law are still necessary. The definition of smoke must be widened [73]; the Alkali Inspectorate should either be strengthened in numbers or (preferably, in the present writer's view) their work should be handed over to local authorities, with the retention of a small centralised research unit; the controls over motor vehicle exhausts should be drastically overhauled and strengthened, and the availability of 'authorised fuels' should be rapidly increased so that many more smoke control orders may be made; also the Minister should be prepared to use his default powers under section 10 of the Clean Air Act, 1968 [74], in any case where a local authority seem to be dragging their feet in the matter of the declaration of smoke control areas.

More general training of young graduates for the public health inspectorate and encouragement by financial inducements to them to specialise in atmospheric pollution control also seems desirable. As in all matters concerned with public health and hygiene, the price of success is 'eternal vigilance', and the country must be willing to pay for an adequate number of sufficiently trained inspectors.

B. INTERNATIONAL ASPECTS

by DAVID HARRIS

Department of Law, University of Nottingham

Introduction

Clean air legislation of the sort that has just been described is the result of internal pressures within the United Kingdom. It is not required by international law, at least not insofar as it is concerned with pollution that has its effect solely within the territory of the United Kingdom. To that large extent, air pollution is regarded by international law as a matter of domestic jurisdiction. Thus, as in the case of legislation about such questions as road safety and cruelty to animals, a state is left to permit or control pollution as it likes. In the future, it might be that customary international law, i.e. the law that is accepted by states as a whole as binding upon them, will recognise a human right according to which everyone is free from air pollution or a rule that a state should at least control air pollution within its territory so far as it affects the nationals of other states, or their property therein. At present, there is little or no evidence to support either development.

International law, therefore, is limited for the moment to the control of what might be called international pollution of the atmosphere, i.e. pollution that arises in the territory of one state and has its effects in the territory of another, on the high seas, or elsewhere. As might be expected from the recent nature of active concern about the problem, the law on such pollution is not well developed. Certain general propositions can, however, be stated concerning both ordinary pollution, resulting from familiar industrial and domestic activities, and ultra-hazardous pollution, resulting from the use of radio-active materials.

The Trail Smelter Rule

Most such propositions, particularly those concerning ordinary pollution, stem from the *Trail Smelter* case [75]. This was a case between the United States and Canada that was referred to arbitration in the 1930s. At the beginning of the century, a Canadian company began smelting lead and zinc at Trail, on the Colombia River about ten miles from the border between the two countries on the Canadian side. In the 1920s, the company stepped up production and by 1930 over 300 tons of sulphur, containing considerable quantities of sulphur dioxide, were being emitted daily. Some of the fumes were being carried down the Colombia River valley and across into the United States where it was allegedly causing considerable damage to land and other interests in the state of Washington. Canada agreed that the case should be referred to the International Joint Commission. This is a body established by the two countries by the Boundary Waters Treaty of 1909 [76] with jurisdiction over problems concerning their common water boundary. It is a notable early example of one kind of standing administrative body that can be of use in controlling international air or water pollution. The Commission reported in 1931 that damage had indeed occurred and assessed it at $350,000. Canada, which had not disputed the question of liability, agreed to pay this amount. But the smelter at Trail continued to operate and the question whether damage had been caused after 1931 for which compensation should be paid also was raised and referred, this time, to arbitration. The United States claimed approximately $2 million compensation. The Tribunal allowed the claim in part, awarding $78,000 compensation in respect of damage caused from 1931 to 1937. The Tribunal also held, in answer to other questions put to it by the parties, that the Trail smelter should refrain from causing further damage and, with the aid of technical experts, laid down an operating regime the observance of which would reduce the emission of fumes to an acceptable level.

What is most important about the case is the formulation by the Tribunal at one point in its award of a rule of international law governing international air pollution. The Tribunal said: '. . . no State has the right to use or permit the use of its territory in such a manner as to cause injury by fumes in or to the territory of another or the properties or persons therein, when the case is of serious consequence and the injury is established by clear and convincing evidence.' [77]

Although supported in the Tribunal's Award almost entirely by prior decisions of the United States Supreme Court on cases involving disputes between states within the United States concerning air and water pollution and not at all by references to the attitude of independent countries in their relations with each other, the rule stated by the Tribunal, so far as it goes, can probably be taken as an accurate statement of international law. It is consistent with the principle *sic utere tuo ut alienum non laedas* (so use your own property as not to injure your neighbour's) that probably applies in the analogous area of river pollution [78]. It can also be based upon the 'neighbourship' principle which, it has been argued [79], applies in international law. It remains the only pronouncement by an international court or tribunal directly in point. Reference is sometimes made to the statement by the International Court of Justice in the *Corfu Channel* case [80] that every state has an obligation 'not to allow knowingly its territory to be used for acts contrary to the rights of other states'.

Yet this rule would not survive in the face of evidence that states generally support a different rule. Neither party to the case, however, has indicated, then or since, its disagreement with the rule and no other state would seem to have done so either. Moreover, what other evidence exists does not support a different rule and is not inconsistent with that spelt out by the Tribunal. Thus, when, in 1961, Mexico complained of smells reaching its territory from stockyards owned privately and located on the United States side of the border, the United States reply was in terms not of absence of liability in international law but of recent steps that had been taken by the companies concerned to bring the situation under control [81].

A number of comments may be made upon the *Trail Smelter* rule. First, it is reminiscent of the rule of common law nuisance in English law which was discussed earlier in this chapter as a means of obtaining a remedy against air pollution in the English courts. As in common law nuisance, liability is based upon unreasonable interference with the enjoyment of land. Secondly, also as in common law nuisance, the interference has to be serious. What would satisfy this test in international law is, to some extent, a matter of conjecture. In English law, factors such as the duration and frequency of the interference are relevant as well as the practice and level of air pollution in the neighbourhood. This last factor may be less helpful to the defendant at the international level if an international standard is applied or if the standards set by the plaintiff state in its territory are taken into account. Examples of what amounts to nuisance in English law are given earlier in this chapter. The rulings by the Tribunal on the factual situation in the case are of some help in indicating the kind and type of interference that creates liability under the *Trail Smelter* rule. Thus compensation was awarded insofar as fumes caused a reduction in crop yield, injury to standing timber and deleterious effects upon the composition of the soil. Although compensation was not awarded in respect of injury to livestock or for loss of resale or rental value of urban real estate, this was the result of failure to prove damage on the facts. The Tribunal would seem to have accepted that, as a matter of law, such claims were within the rule. A claim in respect of injury to business enterprises for 'loss of business and impairment of the value of good will because of the reduced economic status of the residents of the damaged area' was also rejected. The Tribunal stated that 'damage of this nature . . . is too indirect, remote, and uncertain to be appraised and not such for which an indemnity can be awarded' [82].

A third comment upon the rule is that although the *Trail Smelter* case was concerned with sulphur fumes and the Tribunal referred to 'fumes' in its statement of the rule, it is most unlikely that the rule is limited to 'fumes' in any technical sense. It is much more likely that, like common law nuisance, it covers all forms of air pollution. It is probable, for instance, that it would apply if Norway could show that the 'acid rain' that has been reported as periodically falling on its territory causing metal corrosion and injury to fish comes from human activities in some other state.* Fourthly, the rule extends to pollution by private persons as well as by state agencies. In most respects, a state is responsible in international law only for what its agents do. Here, as demonstrated by the facts of the *Trail Smelter* case, states are required to control the activities of private persons on their territory too.

Fifthly, the question arises whether liability under the rule is liability in respect of direct injury to a state—in this case, injury to its territory and to persons and property within its territorial jurisdiction—or in respect of indirect injury in the form of injury to its nationals, or both. The distinction is important in practice in that if the liability is of the former kind there is no need for the claimant state to try and obtain a remedy in the courts of the state allegedly in default before raising the case at an international level; if it is of the latter kind, the individual claiming injury has to go through those courts, which may be time-consuming and expensive before an international claim can be brought. The distinction is also important in that it determines which state may bring the case. If the injury is direct, it is the state whose territory is injured that may do so. If it is indirect, it is the state of which the individual claiming to be injured is a national. Whether the injury is direct or indirect, the claim has to be brought by a state; no individual who claims to be injured has any legal right under the *Trail Smelter* rule or standing to take up the case.

* See *The Times*, 30 November 1970, p. 4. One theory is that sulphuric acid produced by oil-burning factories in the Ruhr and the English Midlands is carried across the North Sea by wind.

So, what kind of liability exists? The Tribunal's wording of the rule is not clear on this point. But it would seem that Canada and the United States purposely presented the *Trail Smelter* case as one of direct injury. One writer closely concerned with the case has written:

'The third difficulty—the technical rules of international law governing the presentation of claims—was overcome by avoiding the format of the ordinary claims convention. The problem presented to the Tribunal for solution was not the problem of damage or injury to this, that or the other person in the State of Washington. It was the problem of damage caused in the State of Washington, regardless of who was hurt. Damage was not related to persons, and the problems of national character of claims and pursuit of national remedies became irrelevant.' [83]

Finally, there is the question of the remedy available where the rule is infringed. In English law there is the possibility of damages and, more significantly in most cases, of an injunction to prevent further injury. Both remedies were, in effect, obtained in the *Trail Smelter* case. In addition to damages, the Tribunal, with the assistance of technical experts, laid down a regime for the future operation of the smelter which would reduce the pollution to an acceptable level.* This was done under a power expressly given to the Tribunal by the parties. It is arguable that it is a power inherent in any international court or tribunal in order to allow it to ensure reparation in appropriate cases.

Pollution from Radio-active Materials

A form of air pollution which requires special consideration is that caused by radio-active materials. This may have both long term and immediate effects. As far as the former is concerned, a gradual

* The following comments by Read on the effects of the regime are of some interest: 'The capital cost of complying with the regime was of the order of twenty million dollars: pre-war costs when a million dollars meant a lot of money. Fortunately for the economic life of the southeastern part of British Columbia, in which the Trail Smelter was the most important factor, it was a private enterprise and not a government department. Notwithstanding the enormous capital expenditure and intermittent interruption of the metallurgical operations imposed, in perpetuity, by the regime of control, the Consolidated Mining and Smelting Company succeeded in selling the products of its smoke abatement programme for substantial profit. In order to comply with the regime, the Company was compelled to remove from the smoke cloud at the stacks more sulphur dioxide than was taken from the stacks of all other smelters of the North American Continent combined.' Ibid., p. 221.

accumulation of radio-active materials in the atmosphere may in the long term reach a level dangerous to world health. It is most unlikely that a state is liable in international law for conduct that could be said to contribute to such a result. Neither the *Trail Smelter* rule nor any other would seem to lead to such liability.

Important steps have, however, been taken by the Nuclear Test Ban Treaty 1963 [84] to prohibit certain activities that increase the amount of radio-active materials in the atmosphere. The Treaty, one purpose of which is 'to put an end to the contamination of man's environment by radio-active substances', bans all 'nuclear explosions', whether for testing nuclear weapons or for peaceful purposes (for example, for civil engineering), in the atmosphere, in outer space and under water. Nuclear tests conducted on the high seas are prohibited. So are tests within the territory of a state from which 'radio-active debris' is carried beyond its territorial limits. In practice this last requirement has not prevented underground nuclear tests. The Treaty does not apply to the use of nuclear weapons in time of war. The legality of such weapons as weapons of war is uncertain [85]. The Treaty is limited to 'nuclear explosions'. It does not apply to other ways (for example, by wastage from nuclear reactors) in which radio-active materials may be knowingly released into the air. Nor does it apply to accidental pollution. Over 100 states have become parties to the Treaty, including three—the United Kingdom, the United States and the USSR—of the five nuclear powers. The other two— mainland China and France—are not parties and are not bound by its provisions. Mainland China conducts its tests on its own territory. They are not in themselves illegal in international law and, in accordance with the suggestion made earlier, it is most unlikely that mainland China is liable in international law for the long term effects of the pollution they cause. The same is probably true of France and its tests on the high seas in the Pacific. It is probable that such tests are lawful as an exercise of freedom of the high seas [86]. The question is now before the International Court of Justice. Australia and New Zealand have brought claims challenging their legality. On 22 June 1973, the court, by eight votes to six, issued provisional measures of prohibition ordering the parties to 'ensure that no action of any kind is taken which might aggravate or extend the dispute submitted to the court or prejudice the rights of the other party in respect of the carrying out of whatever decision the court may render in the case'. The French government, in particular, the court stated 'should avoid nuclear tests causing the deposit of radio-active fall-out on Australian/New Zealand territory'. These measures, which are the normal means of preserving the rights of the parties, in no way indicate how the court will decide the claims—supposing that it first decides that it has the jurisdiction to hear them.

The Nuclear Test Ban Treaty is also directly relevant to immediate injury caused by radio-activity in that it prohibits certain, though not all, conduct that may lead to it. Apart from the Treaty, it is arguable that liability to compensate for immediate injury caused by radio-active fall-out from a 'nuclear explosion' exists under general international law in some situations. The position may be illustrated by reference to the facts of an incident which occurred following a United States nuclear test in the Pacific in 1956 [87]. Japanese fishermen were killed or injured and inhabitants of the strategic trust territory in the Pacific administered by the United States were injured by radio-active fall-out from the test. They were outside the zone that the United States had warned people to avoid. The accident occurred because of a serious miscalculation of the amount of fall-out by United States authorities and because of a sudden change of wind (which may, or may not, have been predictable). The United States gave medical assistance and paid compensation without acknowledging liability in law. The compensation paid to Japan was in respect of personal injuries and economic loss through the contamination of fish. In this incident, the *Trail Smelter* rule, with its roots in the enjoyment of land, would not be applicable. In other cases involving immediate injury from fall-out from nuclear explosions on the territory of an adjoining state it might be. Liability in the 1956 incident might perhaps exist, as far as Japan was concerned, for violation of freedom of the high seas. A state using the high seas for a lawful purpose such as, arguably, nuclear weapon testing, must pay 'reasonable regard to the interests of other states in their exercise of the freedom of the high seas' [88]. Another possible basis for liability, of more general application, is that of strict liability for ultra-hazardous activities, including the use of radio-active materials [89]. There are some indications in treaty practice of such a development [90]. Parallels for such liability exist in municipal law.* It may, however, be too early to regard it as a part of present day international law.

---

* E.g. the Nuclear Installations (Licensing and Insurance) Act, 1959 imposes strict liability in the law of the United Kingdom for the escape of radio-active material.

Development of the Law

As noted earlier in this chapter, in the United Kingdom reliance is placed mostly upon legislation establishing administrative machinery for the control of air pollution, leaving common law nuisance very much as a 'long stop'. Although the latter is helpful in the occasional case where a particular claimant can prove serious damage by a particular defendant or group of defendants and has the inclination, time and money to do so, clearly community interests are best served by continuing public control of pollution setting standards which cannot be infringed except on pain of penalty and with adequate permanent machinery and personnel to ensure their observance.

Although it would be impossible in the present state of organisation of the international community to project this sort of arrangement fully on to the international plane, it is believed that some lessons may be learnt from it. It surely must be the case that international air pollution cannot be adequately controlled just by a rule of nuisance such as that in the *Trail Smelter* case. If a rule of nuisance is, as is believed, inadequate in a national community with compulsory courts, it must be even more obviously so in the international community where compulsory procedures to enforce law are noticeable mostly for their absence. But what further measures can be taken? The best proposal within the present framework of international law is one for a multilateral treaty (or perhaps several applying to particular regions with common problems) which would require states to take adequate steps under their own law to control air pollution. Ideally, the treaty would be aimed at national as well as international air pollution. To such an end it could be conceived in human rights terms [91]. It would allow other contracting states *and individuals* a right of petition to an international body alleging a violation of the treaty by another contracting state. That body would have powers of investigation and decision. Without doubt, the effectiveness of such a treaty would depend upon the strength of its machinery for implementation. Bilateral treaties between neighbouring states dealing with problems of pollution across a common boundary and with adequate enforcement machinery would also be useful, particularly in dealing with pollution problems in the immediate vicinity of land boundaries, such as that in the *Trail Smelter* case [92]. It is interesting to note in this connection the progress that has been made in the similar context of river pollution which is considered below in Chapter 6.

References

A

1. A departmental committee appointed in July 1953 under the chairmanship of Sir Hugh Beaver; it reported in November 1954 (Cmd 9322).
2. The local authorities in England and Wales charged with the administration of the Clean Air Acts are the county boroughs, non-county boroughs, and urban and rural district councils. In London the local authority for the purposes of the Acts is the appropriate London borough: London Government Act, 1963. In Scotland this responsibility is given to the counties and town councils. The local authority, in this sense, is given the duty of enforcing the Acts (1956 Act, s. 29; 1968 Act, 1st Schedule, para. 1) and also is given powers as to research and publicity (s. 25).
3. Formerly the responsible Minister was the Minister of Housing and Local Government; he was replaced in 1970 by the Secretary of State for the Environment (herein called 'the Minister').
4. See p. 73.
5. See p. 75.
6. Although there are separate legal systems and forms of local government in England and Wales and in Scotland, the Clean Air Acts apply to both countries, subject only to minor variations for Scotland in ss. 23(3), 27(5), 28(2), 29(3), 31(7) and 34(1). There are also special provisions relating to Northern Ireland (s. 36).
7. See Beaver Report. One micron is a thousandth part of a millimetre.
8. Ibid.
9. Where a statute provides that a specified word shall 'mean' something for the purposes of that statute, this amounts to providing that the 'something' is to be transcribed into the statute wherever the word defined appears, except where the context makes it quite clear that this definition is not to be applied. The word 'includes' in a definition section of a statute, however, merely extends the meaning of the word so explained without excluding the 'ordinary' meaning or meanings of that word.
10. Maximum penalties are specified in s. 27 of the 1956 Act (as amended by the 1968 Act).
11. Dark Smoke (Permitted Periods) Regulations, 1958 (SI 1958, No. 498).
12. See the Clean Air (Emission of Dark Smoke) (Exemption) Regulations, 1969, SI 1969, No. 1263.
13. 1956 Act, s. 19.
14. Ibid., s. 20. The UK claimed a three-mile territorial sea in 1956, and still does so. Should the Crown, acting under the prerogative, alter its claim the alteration would be read into section 20; see *Post Office* v. *Estuary Radio* [1968] 2 QB 740.
15. Dark Smoke (Permitted Periods) (Vessels) Regulations, 1958 (SI 1958, No. 878).
16. 1968 Act, s. 6; this is subject to certain exceptions; see the Clean Air (Height of Chimneys) (Exemption) Regulations, 1969 (SI 1969, No. 411).
17. 1968 Act, s. 6(4).
18. No such regulations have yet been made.

19. 1956 Act, s. 7.
20. Clean Air (Measurement of Grit and Dust) Regulations, 1968 (SI 1968, No. 431).
21. 1968 Act, s. 2; this section may also be applied to the emission of fumes by Ministerial regulations made under s. 7(1)(a), ibid.
22. See J. F. Garner and R. K. Crow, *Clean Air Law and Practice*, 3rd edition (Shaw & Sons Ltd, 1969).
23. Public Health Act, 1936, s. 101, replacing the Public Health (Smoke Abatement) Act, 1926.
24. Maximum penalty ten shillings.
25. First Schedule, para. 5.
26. In manner provided by s. 285 of the Public Health Act, 1936 (Clean Air Act, 1956, s. 31).
27. In writing, signed by the appropriate officer of the Council (PHA, 1936, ss. 283 and 284).
28. Public Health Act, 1936, s. 93.
29. Ibid., s. 94.
30. Ibid., ss. 95 and 96.
31. 1956 Act, s. 16(2).
32. As amended in points of detail by s. 95 of the Housing Act, 1964, and various provisions of the Clean Air Act, 1968.
33. 1956 Act, s. 11(2).
34. See definition in s. 34(1). The regulations at present in force, listing a number of special kinds of solid fuel and including gas and electricity, are SI 1956, No. 2023, SI 1963, No. 1275, SI 1965, No. 1951, SI 1969, No. 1798 and SI 1970, No. 807, with the title 'Smoke Control Areas (Authorised Fuels) Regulations'. According to National Coal Board practice, the test made before a fuel is 'authorised' by the Department of the Environment, is that there should be less than 20 per cent by weight of volatile matter.
35. Section 11(4), under which the following Smoke Control Areas (Exempted Fireplaces) Regulations have been made: SI 1957, No. 541, SI 1959, No. 1207, SI 1966, No. 217, SI 1969, No. 164, SI 1970, No. 615, and SI 1970, No. 1545.
36. See interpretation of works of 'adaptation' given by s. 14 of the 1956 Act (as amended by s. 95(9) of the Housing Act, 1964).
37. 1956 Act, section 12.
38. In manner provided for by s. 285 of the Public Health Act, 1936; see s. 3(1) of the 1956 Act.
39. In writing and signed by an appropriate officer; PHA, 1936, ss. 283 and 284.
40. As defined in s. 343(1) of the PHA, 1936.
41. 1956 Act, s. 12(2), and Part XII of the PHA, 1936.
42. Ibid., First Schedule (as amended).
43. 1968 Act, s. 8.
44. These are areas more or less arbitrarily chosen by the Ministry as areas where air pollution is heavy and which therefore merit special consideration for the making of smoke control area orders. Sometimes of course the pollution may be due primarily to industrial fumes, which are not controlled under a smoke control area order.
45. At the end of 1968 it was said in the Report of the Ministry of Housing and Local Government for 1967–8 that 804,347 acres of land were then

within approved smoke control area orders and that these covered over four million premises.

46. These grants are naturally popular with householders and are not unattractive to local authorities, for each payment made by them attracts a contribution from the Central Exchequer equivalent to four-sevenths of the amount paid by the local authority (1956 Act, s. 13).

47. Encouraged by the Ministry a local authority, when making a smoke control order, customarily carries out a considerable amount of house-to-house publicity, explaining the effects of the order to occupiers.

48. The production of coke and other specialised fuels by the National Coal Board has not always kept up with demand, and the current increasing use of North Sea gas in place of gas produced from coal, has increased these shortages.

49. Subject to a few exceptions, mentioned below.

50. Registration has to be renewed each year, and a stamp duty (£10 in the case of an alkali work; £6 in other cases) is payable on the certificate of registration.

51. This is done by Ministerial order, now made under s. 11(3) of the 1968 Act. This power has been used sparingly; only six orders have been made, affecting a total of sixty-three premises in Manchester, Leeds, Liverpool, Sheffield and Birmingham.

52. The local authority also may complain to the Alkali Inspectorate of an alleged contravention of the provisions of the 1906 Act; they may then have to pay the expenses of any inquiry held into such a complaint. This procedure is used only very rarely.

53. Made by the Minister under the Public Health Acts, 1936 and 1961; since 1965 the Regulations, applying to the whole country, have replaced the former building byelaws made by local authorities.

54. See p. 69.

55. Public Health Act, 1936, s. 64; if the plans have not been passed within five weeks, approval is to be deemed to have been given; ibid., s. 65(4).

56. Ibid., s. 65(1) and PHA, 1961, s. 4(6).

57. Ibid., s. 65(2) subject to exceptions in s. 64(4).

58. The Road Traffic Act, 1960, as amended by the Road Traffic Act, 1962 and extended by the Road Safety Act, 1967.

59. Motor Vehicles (Construction and Use) Regulations, 1969 (SI 1969, No. 321).

60. This means 'any highway and any other road [i.e. a way which in ordinary use would be described as a road] to which the public has access' [whether as of right or not]: Road Traffic Act, 1960, s. 257(1).

61. Clerk and Lindsell, *Law on Torts*, 13th edition (Sweet & Maxwell, 1969), p. 781.

62. Per Lord Westbury LC, in *St Helens Smelting Co.* v. *Tipping* (1865), 11 HLC 642, at p. 650.

63. Clerk and Lindsell, op. cit., at p. 785, citing Knight-Bruce, VC, in *Walter* v. *Selfe* (1851) 4 De G & Sm 315, at p. 322.

64. *Walter* v. *Selfe, supra.*

65. *Crump* v. *Lambert* (1867) LR 3 Eq 409.

66. *Pwllbach Colliery* v. *Woodman* [1915] AC 634.

67. *Crump* v. *Lambert, supra.*

68. *Sturges* v. *Bridgman* (1879) 11 Ch D 852, at p. 856.

69. *Halsey* v. *Esso Petroleum Co. Ltd* [1961] 2 All ER 145.

70. *Eastern* v. *South African Telegraph Co.* v. *Cape Town Tramways* [1902] AC 381; *Cavey* v. *Ledbitter* (1863) 13 CB NS 470.
71. As in, for example, *Att-Gen.* v. *PYA Quarries Ltd* [1957] 2 QB 169.
72. Public Health Act, 1936, s. 100.
73. The emission of smoke in the United Kingdom was reduced from an average of about 170 microgrammes per cubic metre in 1958 to about 70 microgrammes in 1968. The decrease in sulphur dioxide concentrations over the same period was not nearly so dramatic (see First Report of the Royal Commission on Environmental Pollution).
74. At the time of writing (summer 1971), these powers had not been used in a single case.

B

75. 3 *United Nations Reports of International Arbitral Awards* 1905.
76. 3 Malloy 2607.
77. Ibid., p. 1965. The Tribunal had been instructed to apply 'the law and practice followed in dealing with cognate questions in the United States of America as well as international law and practice'. It found no need to distinguish these two sources because United States law on the relations between states within the United States on air pollution 'whilst more definite, is in conformity with the general rules of international law': ibid., p. 1963.
78. See below, p. 118.
79. See Andrassy, 79 *Hague Recueil* 73 (1951-II). On the doctrine of abuse of rights, which could also be invoked, contrast Oppenheim, *International Law*, Vol. 1, 8th Edition (Longmans, 1955), para. 155a, with Schwarzenberger, 44 *Trans. Grot. Soc.*, 147 (1956).
80. ICJ Reports 1949, p. 4, at p. 22.
81. Whiteman, 6 *Digest of International Law* 256.
82. 3 *United Nations Reports of International Arbitral Awards* 1931.
83. Read, 1 *Canadian Yearbook of International Law* (1963) 213, 224.
84. UKTS 3 (1964), Cmnd 2245; 480 UNTS 43. For a legal discussion of the Treaty, see Schwelb, 58 *AJIL* (1964) 642.
85. See Schwarzenberger, *The Legality of Nuclear Weapons*, 1958, and Falk, 59 *AJIL* (1965) 759.
86. See McDougal and Schlei, 64 *Yale LJ* (1955) 648. For a contrary view, Margolis, ibid., 629.
87. See Whiteman, 4 *Digest of International Law* 565 *et seq.*
88. Article 1, General Convention on the High Seas 1958, UKTS 5 (1963); Cmnd 1929; 450 UNTS 82. The Convention can be taken to state customary international law on this point.
89. See Jenks, 117 *Hague Recueil* 99 (1966-I).
90. See, e.g., the OEEC Convention on Third Party Liability in the Field of Nuclear Energy 1960, UKTS 69 (1968), Cmnd 3755, 55 *AJIL* (1961) 1082, and the Brussels Convention on the Liability of Operators of Nuclear Ships 1962, 57 *AJIL* (1963) 268.
91. In this case, a protocol to an existing human rights treaty such as the European Convention on Human Rights 1950, UKTS 71 (1953), Cmnd 8969, 213 UNTS 221, might be an alternative approach.
92. See, in this connection, Resolution 71(5) of the Committee of Ministers of the Council of Europe.

Chapter 5

Pollution of Inland Waters

by PAULINE K. MARSTRAND
Senior Research Fellow, Science Policy Research Unit, University of Sussex

The Situation

Why is there suddenly so much concern about water pollution? Is it something new? No. Is it worse than it's ever been before? In most countries, again no. What then, has changed? Factors probably contributing to the unrest include: increased mobility, so that more people are appreciating how widespread the problems are. Better communications, so that even unseen problems impinge on the consciousness, and a higher level of scientific understanding so that the possible implications are recognised.

Exacerbating the effects of these factors, the trek to the towns has increased the total quantity of waste accumulation witnessed by any one person in a lifetime, to the point where actual physical discomfort is caused, for example, by foaming or discoloured tapwater, stinking, rat-infested streams, etc. So that whereas a rural community accepts the damage caused to its immediate area, the quality of nuisance caused is far less acceptable when the waste of several contiguous communities is juxtaposed into a seemingly endless zone of destruction.

These developments are peculiar to the already highly industrialised countries. Widespread concern about environmental damage is not a feature of most developing countries. In socialist countries, while scientists are as concerned as those of Britain and the United States have been for two or three decades, there has not apparently been much popular protest. Perhaps this is because the contradiction between increased productivity and an increasing gap in standards between the richest and the poorest is not so blatant. The role of communications media may be critical here. Rapidly growing technology may tend towards alienation, in that it breaks down traditional human relationships faster than new ones can be evolved. If this alienation is combined with the continuous provision of 'scapegoats' by the media, it may tend to produce the kind of hysteria which has characterised recent protest. The protest is genuine and necessary. The abuses are real, but the facts are distorted and the proposed remedies often unrealistic. The result is that the most active protest may result in less-effective measures.

An example is the campaign against phosphate in detergent powders. It has been known for some time that excess phosphate in water is one of the contributory causes of eutrophication, the enrichment of lakes and rivers by plant nutrients, so that excessive weed growth occurs, and enormous populations of microscopic plants (Algae) cause 'blooms' in reservoirs. Scientists in pollution control have been investigating methods of removing phosphate from effluents before discharge, and several technical solutions are now known, though still expensive to operate [1, 2]. Meanwhile, the anti-pollution lobby in the US has succeeded in getting legislation to ban the use of phosphates in detergents. Manufacturers began to substitute nitrilotriacetates [3]. However, the trials so far indicate that these may be far more dangerous to ecosystems and to man than phosphates, and a technology for removing them from effluents is not yet developed. Had the anti-pollution lobby directed its efforts towards enforcing the use of the effluent cleaning technology already available, the final result would have been more satisfactory. For whatever reason, an anti-pollution lobby seems to be a feature of the developed 'West'. Notwithstanding its more fantastic prophesies, if this lobby has succeeded in inclining governments to seek and attend to scientific advice about how to enjoy the benefits of technology while minimising the hazards, then it will be worthy of an honoured place in the history of mankind.

Types and Sources of Pollution

Pollution is here considered to have occurred when a substance having deleterious effects on one or more aspects of water quality is at large in water as a result of human activity.

Substances causing pollution fall into four main categories:

suspended solids,
highly oxidisable organic waste (BOD),
metals and other ions,
soluble organic chemicals.

Some pollutants, for instance iron oxides, fall into more than one category, and all four could be further subdivided, but for present purposes this is not necessary. The activities causing pollution, arranged approximately according to magnitude of effect, are:

1

Disposal of domestic waste

2

Food processing
Glue manufacture
Chemical manufacture, including plastics
Pharmaceutical manufacture
Metal processing

3

Paper making and associated processes
Slaughtering
Wool scouring
Fermentation processes
Soap and detergent production and use

4

Agriculture and horticulture
Forestry
Retting of natural fibres
Wood processing
Bleaching and dyeing
Laundering
Gas and coke manufacture

5

Use and production of radio-active materials [4, 5, 6]

Of these, all, with the possible exception of wool scouring, retting and coal-distillation gas works, are likely to continue, and to become increasingly aggregated, if current demographic trends continue. Populations will increase and will move into higher concentrations in towns. Industry in these towns will increase in amount and variety and will produce more pollution. In countries such as Britain, Czechoslovakia, Germany, Sweden and the United States with advanced chemical industry, problems associated with the processing of traditional fibres are likely to be overtaken by those associated with the production of synthetics. Broadly, this will mean a change from large quantities of process water discharged with high loadings of oxidisable organic material* (BOD), to effluents of lower volume containing much smaller amounts of new and sometimes unidentified organic substances, not necessarily degradeable and frequently toxic to many aquatic organisms. This type of pollutant presents few immediate problems in the treatment of water for public supply, and at present satisfies most control criteria. It does, however, present a possible hazard to the aquatic biota and to human health.

Countries with access to sources of natural gas are likely to change over to the use of this, so pollution by gas liquor containing phenols will diminish, except where plant is kept to supply coke for various industrial uses.

Historically, communities and industries have been sited on the banks of rivers and lakes, and the water has been used to receive and remove all kinds of waste, not only liquid effluents. In older communities, after epidemics of waterborne disease, sewers have usually been laid to rivers and lakes, in order to keep 'black' water out of the streets, surface drains and wells. As communities have grown, this has usually led to an intolerable situation, as in lakes Constance, Erie and others, and legislation has been enacted to deal with pollution 'post hoc' and sometimes too late. Where industry is to be newly sited, however, current knowledge of disposal and pollution problems can be taken into consideration and space allowed for ultimate, if not immediate, on-site treatment of effluents. Because of this historical development, early legislation was directed at controlling health hazards; this was followed by measures to protect fisheries, while procedures to control the quality of water for all potential users have only been introduced in the last two decades in a few countries.

Effects of Pollution

Although only recently a topic of general concern, the ecological effects are the most fundamental, because they are part of a complicated web of relationships which spreads out from the aquatic community to the terrestrial and thence to humans. The spread of pollutants, DDT [7–16], mercury compounds [17], through the world ecosystem is well documented and does not need to be rewritten here. Any kind of pollution causes decreasing diversity of species in the aquatic habitat and this produces a less stable

---

* Measured as BOD, the biochemical oxygen demand, that is the amount of oxygen removed from the water by all processes, mainly biological, during a period of five days in the dark at 20°C.

community in greater danger of rapid deterioration as a result of further change. The ecological changes are at the root of more changes in appearance or suitability for use. If an effluent with high BOD leads to anaerobic conditions, the unpleasant smell and black colour are caused by the elimination of all life except anaerobic bacteria. If a solvent or pesticide kills fish or their food a water will become useless for angling. Excess nitrate and phosphate in domestic sewage or farm waste will cause choking by weeds in ditches and streams and algal 'blooms' (excessive green scum) in reservoirs. Almost all pollutants have ecological effects with consequences reaching far beyond the immediate damage due to appearance or toxicity.

Using Ecological Effects for Detection

These ecological effects could be used to chart the progress of pollution. Any given water is at its most stable when it is supporting a maximum number of different species. This number will depend on the duration of daylight, the type of underlying rock and soil in the collecting area, and the amount and kind of dissolved material. If any deleterious change occurs some of the organisms will be better able to adapt to it than others. Even if none are actually killed, there will be a gradual selection towards the 'resistant' organisms, leading to an alteration in the relative numbers of different species present. By recording populations, it is possible to tell from such changes whether they are suffering deleterious effects, and as long-term records are built up they could be used for assessing the effects of new pollutants; in some cases permanent damaging effects could be avoided if the background information had been assembled.

As long ago as 1908 Kolkwitz and Marsson [18] devised a classification of polluted waters based on Protozoal populations. These 'Saprobien systems' were satisfactory for polluted waters, but did not give enough range of categories for relatively unpolluted waters, or because they were based on organic pollution, did not distinguish sufficiently between different types of grossly polluted water. A further disadvantage was that they required proficiency in the use of microscopical techniques and fairly sophisticated and expensive microscopes. Since that time, first Kolkwitz and Marsson and later other workers [19, 20], attempted to refine the scheme and make it more reliable. Brinkhurst [21] suggested a classification based on algae and Fjerdingstadt [22] one based on sessile benthic communities. This latter is satisfactory in many ways, but again requires skill in microscopy and in

distinguishing between species of micro-organisms such as fila-
mentous algae. In 1964, Woodiwiss [23] described a system using
macroinvertebrates. These are visible without optical aid, and can
be fairly readily assigned to groups after some experience in the
field. Groups abundant in certain situations were selected as 'key'
groups, giving an approximate indication of the quality of water,
and these are then considered in conjunction with the total
number of groups, to give the Biotic Index of the water. Since
coming into use by the Trent and other British river authorities, the
system has been adapted and anomalies reduced and it promises
to become a reasonably rapid way of determining biological
quality without microscopical expertise. Another way of making
rapid assessments without lengthy identification procedures is the
Sequential Comparison Index of Cairns *et al.* 1968 [25]. In this
method the specimens do not have to be identified or sorted into
groups. For each specimen taken from the sample a mark is made
on a record card. Should a run of similar specimens be taken,
the same mark is made for each, but when a different one is taken,
either a new line is started or a barrier mark is inserted between
the two unlike ones, viz.:

either	II	2 alike
	III	3 different from above, but like each other
or	II × III	2 alike. barrier, 3 alike but different from the first 2.

Each of the above illustrates 2 runs with 5 specimens.

The more runs there are for a given number of specimens the
greater the biological diversity. A diversity index is calculated
from

$$\frac{\text{Number of runs}}{\text{Number of specimens}} \quad \text{i.e. } \frac{2}{5} \text{ in the example.}$$

This diversity index can then be associated with levels of pollution
ascertained by chemical analysis. The advantage of biological as
opposed to chemical monitoring is that it is not limited to a
particular time and place. If a damaging substance has been
discharged, the effect on the biota will continue to be noticeable
long after the offending substance has flowed away. The effect
can then be traced upstream to its source and the offending discharge located.

Industries which use large volumes of water must in most cases
discharge them in order to maintain flow in the receiving streams.

The biota of such streams become acclimatised to these effluents and a new equilibrium is established.

Social and Economic Effects

The effects which most concern the interested public are the social effects. Most large bodies of water are a recreational focus for an area, until they become so polluted or choked with rubbish as to be an eyesore. Usually they pass through an interim stage, when they are too bad to be respected, but are still fairly easily retrievable. It is at this stage that dumping of large refuse usually begins and they rapidly deteriorate to a state where restoration would require considerable financial outlay. In many industrial areas the loss of recreational amenity supplied by a river, canal or lake represents the loss of a substantial proportion of the total recreational resource of the neighbourhood, and it is in these areas that protection and improvement of water is most required. Yet these are the very areas in which water is most often neglected.

The financial effects of water pollution are seen in increased costs of water supply and waste treatment, loss of revenue derived from recreational use of water and effects on property values in the neighbourhood of polluted water. Where the costs of treating an effluent before discharge are borne by a manufacturer, this may be reflected in higher prices to the consumer, and it is desirable that an assessment of the amounts involved should be made, in order to get a reasonable distribution of costs. It is probable that the actual cost of treatment on site would be less than the cost of treatment by an external agency [26]. This is because a large part of the cost of treatment is associated with capital equipment, and on-site volumes are considerably less, and also because the biota of the treatment works become 'trained' to degrade a particular complex of substances. In addition, untoward discharges are more likely to be comprehended and notified to the plant operator, so that appropriate steps to minimise the effects can be taken. Plant can be designed to deal with particular types or quantity of effluent, and the treated water can be used again. In some cases reclaimed materials can be used or sold to offset some of the cost of treatment.

Effluent Treatment

Many industries are already treating their effluents in order to reclaim either the water, or the substances in it, or, as in the case of the paper industry, both. These savings in cost are not always passed on to the consumer, so there would seem to be no reason why any increase in costs should be. Pollution control is really the logical development of the process which started with the erection of roofs over and walls around the workplace, and went on to provide dust-extraction and central heating.

Even if the total cost is passed on to the consumer it is arguable that the higher priced 'good' would still represent a better bargain than that obtainable at a lower price with the attendant disadvantages of pollution.

Where the existing arrangements result in a saving in costs to the community when an industry installs plant to treat effluents before discharge to a sewer, there would seem to be a case for the community agreeing to bear some of the cost of the plant, and fiscal arrangements should be designed to facilitate this.

Where the development of a community is closely controlled by one authority, as in British new towns and some socialist countries, it is possible to plan for the most efficient, that is minimum total 'cost', solution to pollution problems. In such circumstances, development need not be allowed until effluent disposal facilities are available, thus avoiding one of the most prevalent causes of pollution in older communities, where development of housing and industry has been allowed to proceed without adequate provision for waste disposal. This is still occurring in areas of new holiday housing and in the suburbs of many cities. The urge to improve the social 'good' derived from industrial enterprise could provide a stimulus to innovation, which, in other circumstances is derived from the need to outstrip competitors or to fill a new market 'niche'.

Where new industrial communities are developing there is no place for the old piecemeal type of development, considering only one advantage at a time. A multidisciplinary team, including representatives of the community involved, must plan the whole development in all its phases. This may seem utopian, but unless it happens, the tensions inherent in the conflict of interests between indigenous and immigrant workers, large- and small-scale enterprise, public and private ownership, rural and urban amenities, will not be resolved. There will be no effective pollution control, and the predictions of the doom-prophets may well be realised. Constraints on personal freedom to exploit all the gadgetry of innovation are required, but they will not work if imposed from above, by whatever form of centralised government.

Personnel

In most countries there is a shortage of suitably trained personnel.

Without personnel, there can be no reliable evidence, without evidence no assessment and without assessment no action. So training is a major priority for action. The immediate requirement is for a combination of courses and in-service training of people in relevant disciplines, to equip them to understand the relationship of their discipline not only to the problems, but to the other disciplines involved. When these people have been made available, research into some of the knowledge required to plan for optimal conditions can begin. Insufficient information is available in the following fields:

(a) Biological data on all inland waters.
(b) methods of monitoring biological quality.
(c) Relative costs of on-site, shared site and municipal treatment for key industrial and agricultural pollutants.
(d) Methods of quantifying social damage due to pollution.
(e) Assessment of health effects associated with various classes of pollutant.
(f) Establishing optimum concentrations and volumes for treatment of various types of discharge to provide design criteria for on-site and shared facilities.
(g) Methods of assessing the advantages—employment, increased production, exploitation of new resources—and weighing them against possible disadvantages—visual disamenity, air/water pollution, noise, concentration of population with increased disposal problems, etc.

Three Responsible Groups

For a programme to operate successfully there must be co-operation between those responsible for industry, those engaged in control, and the public. These three groups each have specific parts to play. Those in industry must know what kinds of substance their processes discharge and in what quantities. They should be prepared either to divulge this information to the control authorities, or to prevent the emission. Because only the individual scientist can know in advance what is likely to be discharged, industry should develop techniques for monitoring new substances and contribute to the costs of testing effluents for unwelcome effects.

The pollution control authorities should monitor all effluents and water resources, carry out testing programmes on effluents and develop robust analytical apparatus and techniques suitable for use in the field.

The public should have access to accurate information about

all kinds of environmental and resource damage, including water pollution, and should exercise respect for water resources during recreational use, and for the conservation laws relating to water. There should be well-advertised methods for the public to communicate with the control authority.

It is possible that regionally based teams of hydro-scientists could undertake much of the responsibility attributed above to the control authority. Such teams could act as back-up forces for front-line pollution control officers attached to river-basins, disposal works or water treatment works and could act as consultants. Financial support would depend on the fiscal traditions of the countries concerned. The German government gives grants to water research institutes which then dispense them to support necessary projects. Soviet Socialist Republics have national Committees for the Protection of Nature, which make proposals to the government, together with the estimated budget for each proposal, and are then allotted funds for all or some of their proposals. Industrial enterprises requiring solutions to problems may assist in supporting investigations, and neighbouring republics may also share the costs of research. In most countries the support could come jointly from central government, local government and industry.

Standards for Effluents

Some countries, for example the United States, USSR, Czechoslovakia, have set standards for a large number of water pollutants, and all have done so for radio-active substances, but most, like Britain, Denmark and France, set standards only for oxygen demand, suspended solids and, in some circumstances, pH, chloride, sulphate and ammoniacal nitrogen. There are arguments for both attitudes. If standards are set, then dischargers can estimate fairly accurately how much it will cost to achieve the standards, and once they have invested in the necessary equipment, can cease to worry about pollution; provided that the equipment is properly maintained and operated. Industrial scientists in Britain have expressed a wish to have standards, and the anti-pollution lobby is also pressing for them. However, once a standard has been set, there is undoubtedly a tendency for all discharges below this level to be considered 'safe'. This has happened in the case of BOD requirements and for radio-active materials. Again, if the standard is achieved by each of several firms, all discharging to the same body of water, the effect on this water could still be catastrophic. For this reason, especially in a country like Britain, with relatively small rivers, many pollution control scientists prefer to have, as at present, powers to set separate consent conditions for each discharge, thus avoiding synergistic effects by considering other discharges to the same water, and other local factors. As the consent conditions are periodically reviewed, this gives more flexibility and with amicable consultation between the parties has resulted in a steady improvement in the condition of Britain's rivers despite the expansion and concentration of many pollution-prone industries, and despite the inadequate financial resources of the river authorities who have had responsibility for administering the control procedures. Those who favour this method of control fear that if standards were set, for example, on allowable levels of organic solvent, dischargers would tend to emit up to this standard, rather than at the lowest achievable rate.

For these reasons there are very real dangers associated with reliance on standards. It would be necessary to provide for frequent review, in order that they should be capable of revision as pollution control technology improves, and there should be some check to avoid the production of toxic conditions by several dischargers all in compliance with the standard, but emitting to the same body of water. Routine biological surveys of such waters might be used to provide an 'alert' when conditions were in danger of deteriorating.

If standards were set, on a national or international basis, it would probably be necessary to offer some financial incentive to encourage dischargers to achieve the lowest possible levels. Charges based on concentration and volume might be imposed, or an agreed pro rata volume charge might be remitted by an amount for every unit reduction in concentration.

Pollution Control Programmes
Probably the implicit aims of pollution control programmes are the same everywhere, namely:

1. To maintain surface and underground water resources in a condition suitable for as many uses as possible.
2. To restore waters which have deteriorated to their potential optimum condition.
3. To maintain adequate supplies of suitable water for domestic, agricultural and industrial use.
4. To provide suitable land drainage and flood control.
5. To provide a sufficiency of water area for recreational use.
6. To maintain and enhance the aesthetic character of areas adjacent to water.

7. To provide a service to operate control procedures.
8. To prepare for future changes in technology, policy, and standard of living.
9. To effect all these without imposing inhibitory constraints on development.

It must be assumed that all concerned, whether in government, industry or community, are alike concerned to conserve and enhance the quality of water and prepared to act in accordance with accurate information and advice. All stand to benefit from rational use of water resources in the long run.

In different countries, different aspects have been emphasised, usually as a result of specific problems which arose, or in response to particular pressure groups. For instance, Azerbaijan developed expertise in dealing with oil pollution, and now carries out research projects for neighbouring Turkmenia, while in Britain, private angling interests achieved legislation which for a long time protected waters fished for salmon and trout, although it made little provision for the protection or improvement of non-angling streams.

In all countries, the operation of any control programme requires certain basic information:

The nature and sources of the main pollutants.
The trends in amounts and frequency, and hence priorities for action.
The present and probable future uses of waters.
Methods of monitoring for present and future pollutants and of removing or inactivating them.
Methods of forecasting the effects on waters of trends in technology in the light of probable future uses; these should include international effects.

Most countries are, or are becoming, conversant with the first three of these and the fourth is the subject of much current research. The fifth category has only just become a subject for investigation.

Initiation of a Pollution Control Programme
In the past, such measures as have been taken have emanated from governments in response to public outcry, usually following some serious incident. Although the USSR did commence industrialisation with the hope of avoiding the worst environmental excesses of industrialisation, the fates of Lake Baikal, the Caspian Sea and Moskva river are now causing concern. The rapid industrialisation

of Japan over the last two decades has led to some of the most spectacular and uncontrolled industrial pollution ever witnessed. Had Japan received its first impetus to technological innovation from a country which already had some constraints upon discharge of harmful substances, this near disaster might have been avoided. It is probably true that so far no country has succeeded in becoming 'developed' without a series of environmental crises.

Perhaps, in the past this was inevitable. Too little was known of the biotic factors to detect the effects at an early stage. Sufficiently sensitive analytical techniques to form the basis of effective monitoring had not been developed, and without computers the information would, in any case, have been impossible to handle. In addition, natural resources seemed limitless and people did not care enough. Now the technical problems have largely been overcome. Control measures can be operated. With improved communications, it is possible for one community to know what has happened elsewhere. It ought to be possible to avoid further disasters of the same kind, even if some unknown new ones must occur at least once.

Any country, developed or not, which proposes to introduce a pollution control programme, needs first a clear statement of aims, in the context of the country concerned, and a statement of the division of responsibilities between various sectors of the community. Secondly, it needs access to the technology necessary to achieve these aims and the ability to train, or to have trained, the technical personnel to use it. Thirdly, it requires adequate legislative backing.

When these requirements have been met, the actual situation must be assessed, together with the potential situations which will arise as populations re-group and as technology is introduced. Bearing in mind the primacy of biological considerations and with the underlying aim of never introducing a development which will not actually enhance the environment for a substantial section of the population and which will diminish it for none, a newly industrialised country might succeed in showing the world how to make technology serve mankind.

In Sweden, since the enactment of the Environmental Protection Law of 1967, it has already been suggested that, in order to preserve sufficient 'natural' environment for the population to enjoy, it may be necessary to set limits on concentration or expansion of urban development, and on industrial growth in some areas. London's 'green belt' policy of the 1930s was such a constraint. For this kind of planning to become a reality it will be necessary for each country to establish multi-disciplinary teams of ecologists, engineers, geologists, demographers, and others, and to develop in these a close understanding of the relationship between the various disciplines and management of the environment. Such teams could develop a truly environmental science.

International Agreements

Except in continental situations, where rivers flow through several countries, inland water pollution is not so crucially in need of international agreement as is air pollution. However, considerations relating to the costs of production in various countries have been raised as an obstacle to pollution control, and it is undoubtedly true that international agreements would help in achieving an all round improvement. A first step might be to establish the procedures and restrictions which are already accepted in more than one country and to find out whether those which have not adopted these could do so. Agreement on some of them, for instance protection of fisheries, might be achieved fairly easily. The European Inland Fisheries Advisory Committee is already working to this end [27]. In other cases, it would be necessary to establish long-term environmental priorities. EEC, FAO, OECD could encourage member countries to adopt similar procedures for assessing the present situation so that accurate comparisons could be made and the distribution of benefits likely to be associated with various programmes could be forecast. This could assist in the allocation of costs involved. Countries which share a watercourse have already in many cases (the Rhine, Saar and Moselle Commissions, Lake Constance Commission and others [28]), accepted a common code relating to monitoring, standards, penalties, and this trend could be encouraged.

The sharing of research expertise and information would enable the pollution budgets of countries with similar problems to be pooled and therefore used more efficiently. This kind of co-operation should not be impossible of achievement, and has already begun in some fields. But to succeed in controlling pollution of water before it is too late, a rigid time schedule is required. 1980 would not be too early for the administrative structure to be completed and 1985 for the programmes to be in full operation. Between 1974 and 1980 each country could be approaching as closely as possible the eventual goal of rational use of available water resources with minimum damage to the environment.

References

1. Robert A. Taft Research Centre, Report No. TWRC-8, 'Alumina columns for selective removal of Phosphorus from wastewater'.
2. F. M. Middleton, 'Advanced treatment of municipal wastes in USA', *Water Pollution Control* (1971), No. 2, pp. 201–11.
3. N. Thom, 'Nitrilo-triacetic acid—a literature survey'. *Water Research*, Vol. 5 (1971), 7, July, pp. 391–401.
4. WHO, *Control of Water Pollution, Survey of Existing Legislation* (1967).
5. J. A. Ternisien, *Les Pollutions et ses effets* and *La lutte contre les Pollutions* (Presse Universitaires, 1968).
6. I. Cheret, *L'Eau* (Editions du Seuil, 1967).
7. N. W. Moore, 'Pesticides and the environment', *J. Appl. Ecol.*, Supplement 3, Cape Holden & Edwards.
8. D. W. Johnson, 'Pesticides and fisheries', *Trans. American Fish Soc.*, Vol. 97 (1968), 4, pp. 398–424.
9. R. L. Rudd, *Pesticides and the Living Landscape* (Faber, 1965).
10. L. R. Bays, 'Pesticide pollution and the effect on the biota of Chew Valley Lake', *Water Treatment and Examination*, Vol. 18 (1969), pp. 295–325.
11. A. V. Holden and R. Lloyd, 'Inland fisheries in Europe—nature and extent of water pollution problems' (1970), FAO, UN, EIFAC 70/SC III-1.
12. F. E. Egler, 'Pesticides in our ecosystem', *Amer. Scientist*, Vol. 52 (1964), pp. 110–36.
13. T. J. Sheets, 'Pesticide build-up in soils', AAAS Pubn No. 85 (1967), Symp. of 133rd Conference Agriculture and Quality of our Environment, 1966.
14. S. Jackson and V. M. Brown, 'Effects of toxic wastes on treatment processes and watercourses', *Water Pollution Control*, No. 3 (1970).
15. OECD, *Pollution of Water by Detergents*, 1964.
16. E. J. Perkins, 'Effects of detergents on the marine environment', *Chemistry and Industry* (3 January 1969), p. 14.
17. 'Mercury in edible fish', *British Medical Journal* (16 January 1971), p. 126.
18. Kolkwitz & Marsson, 'Oekologie der tierschen Saprobien', *Int. Rev. ges Hydrobiol. Hydrogr.*, Vol. 2 (1908), pp. 126–52.
19. Liebmann, *Handbuch des Freshwasser Biologie* (G. Fisher verlag, 1962) (2).
20. Sladecek, *Sci. Pap. Inst. Tech. Prog. Technol. Water*, Vol. 7 (1961), pp. 507–612.
21. R. O. Brinkhurst, 'The biology of the *Tubificidae* in relation to pollution', *Proc. 3rd Seminar on Water Quality Criteria* (Cincinatti, 1957, 1962, 1965).
22. Fjerdingstadt, Review paper in *Int. Rev. Hydrobiology*, Vol. 49 (1964), pp. 63–131.
23. F. S. Woodiwiss, 'A biological system of stream classification', *Chemistry and Industry* (14 March 1964).
24. J. R. Chandler, 'Biological approach to water quality management' *Water Pollution Control*, No. 4 (1970), 1, pp. 415–22.
25. J. Cairns, Abaugh *et al.*, 'Sequential comparison index for non-biologists, *J. Water Pollution Control Fed.*, Vol. 40 (1968), pp. 1607–13.

26. P. N. J. Chipperfield, 'Cost of biological treatment for industrial wastes', *Chemistry and Industry* (6 June 1970), pp. 735–7.
27. European Inland Fisheries Advisory Committee, various reports. FAO, UN, Geneva.
28. Council of Europe, *Freshwater Pollution Control*, 1966.

Chapter 6

The Law Relating to Pollution of Inland Waters

A. IN THE UNITED KINGDOM

by J. GARNER
Professor of Law, University of Nottingham

Introduction

In this chapter we are concerned only with pollution of rivers and natural streams; much of the common law applies also to pollution of canals and other artificial streams, but it is thought that this does not cause many peculiar difficulties in practice. Pollution may be direct, by the discharge or deposit of polluting matter, by seepage from the underground strata, or by run-off and overflow of surface water. For centuries the common law has provided remedies in respect of river pollution without distinguishing between the different forms, but after the industrial revolution and the introduction of piped sewerage in the nineteenth century, it became obvious that modern civilisation required more sophisticated controls and government agencies charged with the duty of administering these controls. The common law was not, however, replaced by these statutory controls, and the two systems continue to exist, side by side, the common law having the advantage of a much more stringent definition of pollution than has as yet been adopted for the statutory controls, but also having the serious disadvantage of dependence on the initiative of some individual (normally a riparian landowner) and being subject to the uncertainties and cost of civil litigation.

Much of the pollution in English rivers is caused by sewage to rivers, but the discharge of untreated sewage to rivers has been decreasing substantially in volume in recent years. Industrial discharges are still responsible for much pollution, and where these exist they tend to be of a persistent kind. The commonest polluting industries are leather and hide, chemicals (especially phenol), metallurgy, textiles (cleaning), gas and coke, mining and paper. In more recent years a new source of pollution has become apparent, namely that caused by the use of pesticides and similar substances on land in the vicinity of a river, which then reach the stream by overflow or borne by surface water. The several statutory controls will be considered separately but first we must consider the common law.

The Common Law

It is a 'natural' incident [1] of the rights of ownership of land abutting on a river or natural stream that the owner should be entitled to the flow of water in the stream in its natural state, undiminished (and not exceeded) in quantity, and unpolluted by artificial liquid and solid matter in suspension or in solution, and of the physical (e.g. as to temperature [2]) or chemical consistency of the water in its natural state. The only exceptions to this principle are cases *de minimis*, such, as for example, where the water was temporarily made muddy [3], and also cases where a right to pollute a stream (to a defined extent) has been acquired by an express grant or (more commonly) by a continued amount for a full period of at least twenty years [4]. In a leading case in the House of Lords [5], Lord Macnaghten said,

'A riparian proprietor is entitled to have the water of the stream, on the banks of which his property lies, flow down as it has been accustomed to flow down to his property, subject to the ordinary use of the flowing water by upper proprietors, and to such further use, if any, on their part in connection with their property as may be reasonable under the circumstances. Every riparian proprietor is thus entitled to the water of his stream, in its natural flow, without sensible diminution or increase and without sensible alteration in its character or quality. Any invasion of this right causing actual damage or calculated to form a claim which may ripen into an adverse right, entitles the party injured to the intervention of the Court.'

The intervention of the Court may take the form of an award of damages, where some injury to the plaintiff's rights can be established, or an injunction where it is desired to prevent an adverse claim being established by prescription; in appropriate circumstances both remedies may be granted in the same proceedings. Only a riparian landowner may normally sue, but if the pollution is so grave as to amount to a public nuisance the defendant may be prosecuted on indictment, or civil proceedings for an injunction may be commenced by the Attorney-General 'at the relation of' a party affected, or possibly of the local authority [6].

In proceedings for damages or an injunction the riparian landowner as plaintiff will have to prove actual pollution to some extent, although he need not establish actual injury, but these proceedings may be taken only by some person entitled to the flow of
water and not (for example) by a plaintiff who was *de facto* taking
the water without legal authority [7].

These principles apply also to the fouling of percolating water
under a plaintiff's land (which he will normally *not* own), even if
such water does not run in a defined channel. As was said by
Brett, M.R., in a leading case [8],

'Although nobody has any property in the common source [the
reservoir of percolating water under the plaintiff's land] yet every-
body has a right to appropriate it, and to appropriate it in its
natural state, and no one of those who have a right to appropriate
it has a right to contaminate that source so as to prevent his neighbour from having the full value of his right of appropriation.'

Thus, Mrs Hubbs, who had complained 'that her water was not
potable to the District Engineer' [*sic!*], was entitled to sue defendants who had dumped a substantial quantity of sand containing sodium and calcium chloride on their land about 100 feet
away from Mrs Hubbs' well, from which she drew her drinking
water [9].

Liability under this branch of the law may also attach to a defendant who deliberately or without due care allows some noxious
substance, such as tar or a pesticide, to escape from his land and
so pass directly or indirectly into the waters of some stream or an
artificial conduit, and thereby cause injury to the rights of the
plaintiff. Thus, the owner of some watercress beds was awarded
damages when his watercress was ruined by tar flowing from the
defendant local authority's highway [10], and similar results
followed when fumes from creosoted wood blocks damaged the
crops of a market gardener [11].

In all these cases it is no defence to an action alleging pollution
of a stream to show that the stream is already polluted from some
other source; 'what matters is whether what is added would appre-
ciably pollute the stream if its water were otherwise pure' [12].
A right to pollute may, as we have said, be acquired by prescription (subject to the statutory provisions described below); but a
prescriptive right may not be exceeded: 'a proprietor who has
prescribed a right to pollute cannot in my opinion use even his
common law rights in such a way as to add to pollution' [13].

In spite of the twentieth-century legislation on the subject, these common law provisions are still used on occasion, although of course they can be used only by a plaintiff riparian owner; and not all riparian owners are concerned about the condition of the river or stream passing their land, and even those who are concerned may have neither the pertinacity nor the financial resources to take legal proceedings in respect of infringements of their rights. The outstanding example of such proceedings being taken by private individuals in recent years was *Pride of Derby and Derbyshire Angling Assn Ltd* v. *British Celanese Ltd* [14]. In this case, an action was brought by an angling association, the owner of a right of fishery [15] on the rivers Derwent and Trent, and the Earl of Harrington, a riparian landowner of a considerable stretch of both rivers. The defendants were Derby Corporation, who discharged insufficiently treated sewage matter into the river from their sewage works; British Celanese Ltd, who discharged a heated effluent containing suspended organic matter; and also the British Electricity Authority, who discharged large quantities of heated effluent from their generating station, the water in the latter case having been taken from the river. Injunctions were granted against all three defendants, but that against the Electricity Authority was confined to restraining them from returning the water after use in such a condition as to cause injury to fish.

Normally pollution means the addition of some substance to the water, or possibly the heating of the temperature of the water, but the common law will also intervene if a defendant substantially diminishes the flow of water passing the land of a riparian owner lower down the stream, unless the defendant takes the water solely for 'natural' purposes (i.e. for such purposes as the watering of his cattle or for domestic purposes). The taking of water for the spray irrigation of the defendant's crops is not a natural purpose in this sense, and therefore, as use of water for this purpose results in only a very small quantity of the water returning to the river, such extraction was held to be actionable at the suit of the plaintiff riparian owner lower down the stream [16]. It is, however, a defence in common law proceedings arising out of the abstraction of water if it can be shown that such abstraction was made pursuant to a licence granted by the river authority under section 31 of the Water Resources Act, 1963.

It is now proposed to turn from the common law to statute, which latter must be considered under two main headings, namely legislation controlling direct pollution of a river, stream or other waters, and secondly legislation controlling the use on land of pesticides or other noxious substances which may, by spillage or percolation, directly or indirectly get into a river or stream by no precisely definable course.

Statutory Controls: River Pollution

Parliament has concerned itself with the prevention of river pollution since the last quarter of the nineteenth century [17], but the law is now contained in the Rivers (Prevention of Pollution) Acts, 1951 and 1961 (as amended and supplemented by the Water Resources Act, 1963), and also the Salmon and Freshwater Fisheries Acts, the Radio-active Substances Act, 1960, and the Oil in Navigable Waters Acts. It should first, however, be made clear that the responsible government agency for administering most of the river pollution legislation is now the river authority [18], of which there are twenty-nine, one for each of the principal watersheds in the country, and will shortly be the regional water authority. The legislation deals with a number of different aspects of pollution:

(a) the pollution of streams and watercourses other than by 'deliberate' discharges of effluent;
(b) discharges of effluents to streams and watercourses, of either sewage or industrial effluent;
(c) discharges to sewers of sewage or industrial effluent (which then themselves discharge, with or without treatment, to streams, etc. as in (b));
(d) discharges to percolating waters;
(e) discharges to tidal waters (discharges to the sea directly are considered to be outside the scope of this chapter);
(f) 'fishery' provisions;
(g) the discharge of radio-active substances, and other miscellaneous provisions.

These several headings will now be considered *seriatim*.

(a) *Pollution of streams, etc.*   The principal provision under this heading (but see also para. (f) below) is section 2 of the Rivers (Prevention of Pollution) Act, 1951. This section provides that it shall be an offence if any person causes or knowingly permits to enter a stream [19] any poisonous, noxious or polluting matter, *or* any matter so as to tend either directly or in combination with similar acts (whether his own or another's) to impede the proper flow of the water in the stream in a manner leading or likely to lead to a substantial aggravation of pollution due to other causes or of its consequences. There is thus no offence under this section if the

person responsible is not aware that his effluent is discharging
into a stream [20]. It is also not clear what is meant by the words
'poisonous, noxious', etc.; do they include poisonous to fish
generally or a particular kind of fish, or is an effluent that is merely
harmful to some kind of fish life within the section? Again, is
'polluting' to be understood in the wide common law sense, and
is the word 'matter' to be read as a scientist would, so as to include
solids, liquids and gases? These and other difficulties of interpretation of this section have not yet been resolved by the courts.
It is, however, clear that the fact that a defendant may have been
discharging the matter complained of over a long period of time
is no defence; a prescriptive right to pollute may be acquired at
common law, but not under this section.

Incidentally, it is also provided in this section that it shall not be
an offence if water which has been raised or drained from any
underground part of a mine is discharged into a stream in the same
condition in which it was raised or drained from underground.
There is also a general provision in section 38 of the Countryside
Act, 1968 to the effect that in the exercise of their functions under
that Act it is the duty of the Countryside Commission, the Forestry
Commission and local authorities, to have regard to protection
against pollution of any water belonging to statutory water undertakers; it is unfortunate that this was not drafted more widely so
as to apply to any water in its natural state.

(b) *The discharge of effluents.* In each of the following circumstances the consent of the river authority must be obtained:

  (i) the making of a new or altered outlet for discharge of trade
      or sewage effluent to a stream [21];
 (ii) the making of any new discharge of such effluent, whether
      or not by means of an existing outlet [22]; *and*
(iii) the continuance of an existing discharge of such effluent
      being made in 1951 [23];
 (iv) abstraction of water.

In each case, it is made an offence for such an operation to be
carried out without the consent of the river authority, or in breach
of any conditions imposed by the river authority in granting such
consent, and the river authority is required to maintain a register
(open to public inspection) of any such consents and conditions
[24]. Any such consent may not be unreasonably withheld and an
appeal lies to the Minister against any refusal of consent.

Each case has to be considered on its merits by the river

authority, there being no standards prescribed in the statute as to when a discharge should or should not be permitted, or as to the conditions to be imposed. In practice it is obviously desirable that the authority should have as wide a discretion as possible, and authorities have used their powers so as to secure gradual improvements in the level of pollution in their rivers without imposing too heavy financial burdens on sewerage authorities and industrialists. Potentially, however, a river authority may stop the discharge of any trade or sewage effluent whatever.

(c) *Discharges to sewers.* Most sewers belong to the local authority [25], but the owners of premises within the district of a local authority [26] have a right to cause their private sewers and drains to communicate with any public sewer and to discharge domestic sewage thereto [27]. The local authority has duties to cleanse, maintain and empty such sewers [28], and to provide sufficient sewers for the drainage of its district [29]; clearly the sewers must lead somewhere, and the natural outfall, before or after treatment, will be to a river or directly to the sea. The duties of the local authority on the one hand to provide an efficient sewerage system and of the river authority on the other hand to prevent river pollution, may clearly conflict on occasion; this is one of the reasons why the Acts of 1951 and 1961 have left a discretion with the river authorities and have not imposed a total prohibition on the discharge to rivers of effluents from local authority sewerage systems.

In the case of industrial ('trade' [30]) effluents, these may not be discharged to a public sewer except pursuant to a trade effluent agreement made with the local authority, or after a trade effluent notice has been served on the local authority [31], and then only in accordance with the terms and conditions (which may be varied from time to time [32]) of any consent issued by the authority. These conditions may regulate such matters as the rate of flow of the effluent, the elimination or diminution of any specified constituents of the effluent, and its temperature, etc., and the local authority may in some cases have to treat the effluent specially before it is discharged to a river. In such (and other cases) the authority may make a charge as a condition of its consent.

(d) *Discharges to percolating water.* It is prohibited to discharge any trade effluents or other poisonous, noxious or polluting matter by means of wells, boreholes or pipes, into any underground strata [33] within the area of a river authority, except with the consent of that authority [34]. This is well enough so far as it goes, but is subject to the obvious limitations that a discharge may be made otherwise than by one of the three methods stated, and of course the point of discharge may be outside the area of the complainant, or indeed any, river authority.

(e) *Discharges to tidal waters.* The controls over discharges to streams outlined in sub-para. (a) above apply generally also to discharges to tidal waters, in so far as these have been specified for the purpose in orders made by the Minister; the control of new discharges under section 7 of the 1951 Act apply to tidal waters within the seaward limits specified in the Schedule to the Clean Rivers (Estuaries and Tidal Waters) Act, 1960 [35], and are there referred to as 'controlled waters'. The limits of controlled waters may be extended by Ministerial order made under the 1960 Act.

(f) *'Fishery' provisions.* Under the Salmon and Freshwater Fisheries Acts, 1923-65, which are enforceable not by local authorities but by the river authorities, it is made an offence to cause or knowingly permit to flow or put [36] into waters containing fish, or into any tributaries thereof, 'any liquid or solid [but not, apparently gaseous] matter to such an extent as to cause the waters to be poisonous or injurious to fish or the spawning grounds, spawn or foods of fish', unless this was being done before the 1923 Act came into force, and provided the best practicable means within a reasonable cost have been used so as to prevent injury to fish [37].

Once again, no definition is given of the word 'poisonous', etc.: would for example some matter which affected the breeding habits of fish be included, and must the poisonous matter affect the life of the fish, or would it apply also to a case where the fish were rendered poisonous as food for human beings? 'Fish' itself also is not defined, but the expression presumably includes shellfish and winkles. Unlike the Rivers Prevention of Pollution Acts, a prescriptive right may be acquired to contravene this section, the ordinary law of prescription being expressly preserved.

It is also provided that no person may use in or near any waters any explosive substance, any poison or other noxious substance, or any electrical device, 'with intent thereby to take or destroy fish' [38].

These provisions are of course designed to protect the fish in a river or stream, and are only incidentally designed to prohibit pollution; the addition of substances which are harmless to fish but spoil the amenities of a river would not contravene the present sections.

(g) *The discharge of radio-active substances, and other miscellaneous provisions.*

(i) *Cemeteries.* Offensive matter from a cemetery may not be caused or suffered to be brought or flowed into a stream in any manner whereby the water is fouled; and in the event of a breach of this section the local authority or cemetery company responsible will be liable to prosecution [39].

(ii) *Gas works.* A gas board or other person engaged in the manufacture of gas may not cause or suffer to be brought or to flow into any stream, etc., any 'washing or other substance produced in making or supplying gas', and also may not 'wilfully' do any act connected with the making or supply of gas whereby the water in a stream, etc., is fouled [40]. There are similar offences contained in the Gas Act, 1948 [41], and a gas board will also be guilty of an offence if washings from gas works are allowed to foul waters belonging to water supply undertakers [42].

'Gas' for the purposes of these provisions is not defined, but is commonly understood to refer to coal or oil gas supplied by undertakers as a source of power or energy to domestic and industrial, etc., consumers.

(iii) *Alkali works.* Waste from an 'alkali' works [43] may not be deposited or discharged without the best practicable means being used effectually to prevent any nuisance [44]; this provision clearly, but not expressly, includes the discharge of such waste to a stream.

(iv) *Animal carcases.* The carcase of an animal which has died of disease or has been slaughtered as diseased or suspected of being diseased, may not be thrown or placed into or in any river or stream, etc. [45].

(v) *Water supplies.* The pollution of water supplies belonging to undertakers is an offence under the Water Act [46], and undertakers may, and in practice do, make byelaws for the protection of their supplies [47], and also may acquire land for the purpose [48].

(vi) *Radio-active substances.* The Radio-active Substances Act, 1960 makes it unlawful for any person carrying on an 'undertaking', by way of trade, business or profession, to dispose of or accumulate radio-active waste material [49], on or from premises used for the undertaking except in accordance with an authorisation granted by the Minister [50], unless the premises are used by the Atomic Energy Authority or is a site licensed under the Nuclear Installations (Licensing and Insurance) Act, 1959 [51]. This control so centralised in the Secretary of State replaces the 'pollution' provisions considered earlier in this chapter so far as radio-active substances are concerned [52], and therefore (for example) a river authority could not take proceedings in respect of an alleged pollution of water in a river by the discharge of radio-active material from an undertaking; the appropriate action in such a case would be for the river authority to complain to the Secretary of State, leaving it to him, if he saw fit, to take proceedings against the offending undertaking under the Radio-active Substances Act. The statute does however impose the further requirement that in any case where disposal of radio-active waste is likely to involve the need for special precautions to be taken by a public authority (such as a river authority) the Secretary of State must consult with that authority before granting an authorisation for such disposal [53]. In practice, such consultation is readily undertaken, but it may not always be appreciated in advance that special precautions may prove necessary.

Statutory Controls: the Use of Pesticides, etc.

The law regulating the use of pesticides and other chemical agents on land is unco-ordinated and fragmentary. Clearly, if a can containing the residue of some toxic agent is thrown into a river, there will be an offence under section 2 of the Rivers (Prevention of Pollution) Act, 1951 [54], but it may not always be possible to identify the culprit, and there are many other circumstances in which these highly dangerous substances may reach the waters of a river or stream [55]. The best cure for such cases is prevention, but the law is at present a clumsy and inadequate instrument. Such statutory provisions as there are will be summarised briefly.

These provisions have been drafted (like the 'fishery' provisions, above) with objects other than the prevention of pollution or the protection of the environment, and this is of course the main reason for their inadequacies. The principal provision is that contained in section 8 of the Protection of Animals Act, 1911 [56], which makes it an offence to place on any land any fluid or edible matter which has been rendered 'poisonous' (a term which is not defined) [57] *except*:

    (i) if the substance is used for the purpose of destroying insects and other invertebrates, or rats, mice or other ground vermin where this is found necessary in the interests of public health, agriculture or the preservation of other animals, whether domestic or wild: *or*

(ii) where the substance is used for manuring the land; *or*

(iii) where a licence has been obtained from the Home Secretary or the Natural Environment Council for the purpose of killing or taking wild birds.

When it is appreciated that it is the duty of the occupier of land to destroy certain pests, such as Colorado beetles [58], and that listed poisons may readily be obtained by a purchaser engaged in the trade or business of agriculture or horticulture [59] it will also be appreciated that this provision is not capable of achieving much in the matter of the prevention of pollution. Moreover, the gassing of rabbit holes is permitted, so as to reduce the number of rabbits [60]. The Ministry of Agriculture, Fisheries and Food gives extensive advice to farmers on the use of pesticides [61] but voluminous literature cannot always be studied in detail by overworked and understaffed farmers, and it would seem preferable that the use of noxious pesticides should be controlled by regulations having legal effect, as indeed has recently been recommended by the Committee of Ministers of the Council of Europe [62].

Conclusion

The statutory controls depend primarily on the initiative of the river authorities; potentially the authorities have most if not all the legal powers they need, but a zealous use of these powers is often impracticable on the ground of the expense that this would cause to the industrialists and local authorities responsible for discharges to the river. It has been recently suggested that river authorities should have wider responsibilities, so that they would become the sewage disposal authorities (in place of the local authority) and also the water supply authority. This has now been accepted by the Government and the proposals are contained in the Water Bill now (summer 1973) before Parliament. It is intended the administrative changes will take place on 1 April 1974. There will thus be one regional water authority concerned with the use, maintenance and disposal of water in the area of a watershed, which would have obvious advantages. On the other hand it is arguable that there would be a conflict of interests between the sewage disposal function and the river pollution prevention function; moreover, an agency with such wide powers should be in some measure responsible to an electorate, and not be as politically independent and remote as are the present river authorities [63].

Yet another suggestion, put forward by the Water Resources Board in their annual report for 1970, was that there should be a national water authority. This would have cut across all geographical watersheds and would have been given as its principal functions the whole field of water management, have been responsible for planning and co-ordinating both the supply of water and the disposal of effluent; it would have been responsible for the development and operation of regional water conservation schemes. Such a national body would have left with existing local authorities most of their sewerage functions (except disposal), and the water supply undertakings could retain their separate existence [64]. As with every national undertaking of this scale there would have remained the problem of control by Parliament.

The common law provides a useful if expensive 'longstop' to the existing statutory controls, available to a particular riparian landowner where the public agencies are not prepared to act. In the more serious cases of pollution by pesticides, the common law may prove of more value than the statutory controls [65]; that this should be so, is only further evidence that statutory controls over the use of pesticides should be considerably strengthened. In particular, their enforcement should be made the specific responsibility of some government agency with adequate resources.

B. INTERNATIONAL ASPECTS

by DAVID HARRIS

Department of Law, University of Nottingham

Introduction

The international law on inland water [66] pollution, like that on air pollution, is not well developed. Here too the relatively recent development of awareness of the problem and the seriousness of its implications, coupled, in the case of river pollution particularly, with thinking that emphasises state sovereignty, have so far meant the absence of a settled and detailed legal regime on the subject. Unlike air pollution, however, river pollution has the advantage that states have long adopted the practice of making treaties laying down rules for the use of particular international rivers. Occasionally, provisions concerning pollution have been inserted in treaties establishing legal regimes governing the use of a river generally. More recently treaties dealing exclusively with pollution have been adopted or proposed. In this respect, control of river pollution, with its greater immediate economic consequences, is more advanced in international law than that of air pollution.

Customary International Law

Before looking at the relevant treaty law, which applies only to the states that are parties to it, it is necessary to consider the position in customary international law, i.e. in the law that applies to states generally, whether parties to any treaties or not.

In customary international law, the first distinction to make is between national and international rivers. Pollution of national rivers, i.e. rivers which flow only through the territory of one state, is at present a matter of domestic jurisdiction and hence not regulated by international law. There are, therefore, no remedies for the pollution of such rivers other than those that may be provided by the state concerned in its law at its discretion. The position in fact is the same as it is in the case of national air pollution. It would be different if a human right to be free from pollution were, in the future, to be developed or, as far as nationals of other states alone are concerned, in the unlikely event that it were to come to be recognised that a state's undoubted duty in international law to treat aliens with a certain degree of respect requires the non-pollution of rivers that they might want to use. Pollution of national rivers may, however, indirectly be subject to international control in a few situations. There may be liability if polluted water percolates underground into the territory of another state. It may also exist if polluted internal waters enter the territorial sea of a neighbouring state or the high seas.*

With regard to international rivers, i.e. rivers that flow through the territory of more than one state, an early view, which followed from the idea of state sovereignty, was that there are no restrictions in international law upon a state's freedom to use them as they flow through its territory. This view, which became known as the Harmon Doctrine, was taken by the United States at the turn of the century in disputes with Mexico and Canada concerning its diversion of the Rio Grande [67] and the Chicago rivers [68] respectively. Clearly, the Doctrine could be applied to river pollution too. In each case, the other country concerned denied its validity. More recently, the United States would seem to have abandoned it. In 1962, its representative to the General Assembly of the United Nations stated: '... In the absence of specific treaty provisions to the contrary, the trend of the law is that no state may claim to use the waters of an international river in such a way as to cause material injury to the interests of other states, nor may a state oppose use of river waters by other states unless this causes material injury to itself...' [69]. The Doctrine has received only limited support from other states and in two recent disputes, concerning the River Jordan in the Middle East and the River Lauca in South America, the upper riparian state did not rely upon it [70].

An alternative viewpoint is that international law does limit the use of an international river by each riparian state in the interest of each of the others, although precisely how strict the rule is remains uncertain. The recent tendency is away from a rule of absolute liability and towards a rule based upon reasonable use. Support for some degree of limitation can be found in what few cases there are and in the views of writers and international groups of lawyers.

The rule stated in the *Trail Smelter* case, although concerned directly with air pollution, and hence considered in detail in Chapter 4, could easily be extended to cover river pollution too. Certainly the principle *sic utere tuo ut alienum non laedas* (so use

---

* See below, Chapter 9.

your own property as not to injure your neighbours) which under-lies it is as applicable to river pollution as it is to pollution of the air. There is also the *Lake Lanoux* Arbitration [71] which was decided in 1957. In that case Spain complained that France had violated a treaty by diverting a river in French territory before it entered Spain. The Tribunal found no violation of the Treaty because Spain could not show that the effect of the diversion had been detrimental to it in any way. In considering on what sort of ground Spain could have had a good claim based upon the diversion, the Tribunal stated: 'it could have been argued that the works would bring about an ultimate pollution of the waters of the Carol or that the returned waters would have a chemical compo-sition or a temperature or some characteristic which could injure Spanish interests' [72]. This gives general support to the view that pollution of an international river may give rise to liability. In addition to these two international decisions, there is the Italian case of *Société Énergie Électrique du Littoral Méditerranéen* v. *Compagnia Imprese Elettriche Liguri* [73] in 1939. In this case, which was also one of water diversion, the Italian Court of Cas-sation stated:

'International law recognises the right on the part of every riparian state to enjoy, as a participant of a kind of partnership created by the river, all the advantages deriving from it for the purpose of securing the welfare and the economic and civil progress of the nation. . . . However, although a state, in the exercise of its right of sovereignty, may subject public rivers to whatever regime it deems best, it cannot disregard the international duty not to impede or to destroy, as a result of this regime, the opportunity of the other states to avail themselves of the flow of water for their own national needs.' [74]

Again, this is a general statement that could apply to pollution as well as to water diversion.

Most, though not all, writers support a limitation upon the rights of riparian states. Oppenheim, for example, states:

'. . . it is a rule of International Law that no state is allowed to alter the natural conditions of its own territory to the disadvantage of the natural conditions of the territory of a neighbouring state. For this reason a state is not only forbidden to stop or divert the flow of a river which runs from its own to a neighbouring state, but likewise to make such use of the water of the river as either causes danger to the neighbouring state or prevents it from making proper use of the flow of the river on its part.' [75]

Similarly, international organisations of lawyers have rejected the Harmon Doctrine. As early as 1911, the Institute of International Law resolved that 'any contamination of water, by means of the discharge therein of injurious matter, is forbidden' [76]. More recently, the International Law Association has stated certain more detailed and, in keeping with current trends, less absolute rules in its 1966 Helsinki Rules on the Uses of International Rivers [77]. These were understood by the Association, which, like the Institute of International Law, consists of individual lawyers and not representatives of governments, to reflect customary international law. The Rules constitute a comprehensive legal regime for 'international drainage basins'. By this term is meant 'a geographic area extending over two or more states determined by the watershed limits of the system of waters, including surface and underground waters flowing into a common terminus' [78]. The regime includes, among other things, rules on pollution. Article IX defines 'water pollution' as 'any detrimental change resulting from human conduct in the natural composition, content, or quality of the waters'. Article X then states the following rule:

'1. Consistent with the principle of equitable utilisation of the waters of an international drainage basin, a state

(a) must prevent any new form of water pollution or any increase in the degree of existing water pollution in an international drainage basin which would cause substantial injury in the territory of a co-basin state, and
(b) should take all reasonable measures to abate existing water pollution in an international drainage basin to such an extent that no substantial damage is caused in the territory of a co-basin state.'

The meaning of 'substantial injury' is indicated in the Commentary to the Rules which reads: 'Generally, an injury is considered "substantial" if it materially interferes with or prevents a reasonable use of the water' [79]. The obligation to prevent 'substantial injury' is not an absolute one; a state causing it may be saved by the 'principle of equitable utilisation'. This principle, which underlies the Rules as a whole, is that each 'basin state is entitled, within its territory, to a reasonable and equitable share in the beneficial uses of the waters of an international drainage basin' [80]. Its object is 'to provide the maximum benefit to each basin state from the uses of the waters with the minimum detriment to each' [81]. In the context of pollution, a riparian state may be acting lawfully in accordance with the Rules even though it causes 'substantial injury' to another riparian state by pollution arising in its territory if it can show that the acts causing the pollution are acts that it may take in realisation of its equitable share of the use of the resources of the river. The Commentary to the Rules gives the following example:

'An irrigation district, an agency of State A, initiates an extensive reclamation and agricultural irrigation project within its territory utilising the waters of international drainage basin X, which flows into the territory of State B. As a result of the irrigation process, the salinity of international drainage basin X is increased to such an extent that it substantially injures agriculture in State B, where the waters of X are used for irrigation. State A is under a duty to prevent injury in State B if its irrigation district is inconsistent with the principle of equitable utilisation of drainage basin X.' [82]

As stated earlier, the Helsinki Rules were thought by their drafters to represent present customary international law. They certainly reflect the trend away from the Harmon Doctrine and towards a more co-operative and equitable approach. They are also realistic in that they do not insist upon total abatement of existing pollution. Yet despite their sophistication, the generality of the terms in which they are couched presents a problem in that there are no institutions to give them meaning in particular cases. Nor do they offer any administrative machinery with powers that would help prevent pollution before it is too late; all that they provide is a prohibition, coupled with a right to obtain an injunction and to recover damages [83]. As in the case of air pollution,* it seems likely that some form of machinery for continuing public control of water pollution is needed in addition to rules of the common law nuisance sort. This can only be provided by treaties.

Treaties
Customary international law is supplemented by an increasing number of treaty provisions on the pollution of particular rivers. These vary in their content and it is probably impossible to extract from them support for any rule that could be said to be one that applies to countries as a whole and not only to the parties to them. It is at least arguable, however, that river pollution is, in any event, a problem that needs to be dealt with on a case-by-case,

---

* See above, Chapter 4B.

rather than a universal, basis and that the best means of tackling it is, in fact, by the adoption of more and more treaties between the riparian states of particular rivers.

The earliest examples of treaties limiting pollution are treaties concerned wholly or partly with fishing which limit pollution in the interests of conserving fish stocks. This, incidentally, is one context in which a lower riparian state may be under an obligation concerning pollution towards a state higher up the river; normally, in the nature of things, it is the other way about. An early example of a more general provision is the Boundary Waters Treaty of 1909 [84] between Canada and the United States which reads (Article IV): 'It is further agreed that the waters herein defined as boundary waters and waters flowing across the boundary shall not be polluted on either side to the injury of health or property on the other.' Exactly how a tribunal would interpret such an unlimited prohibition so as to deal with questions of fault (what if all reasonable care to prevent pollution has been taken?), acquiescence (what if the injured state has known about and failed to protest at the pollution for some time?) and the degree of injury needed to establish liability (is any injury, however small, sufficient?) is not clear. As to this last point, it would probably be possible in a treaty between two countries concerning particular waters in the same environment to set a precise, scientific level at which pollution may be said to be actionable.

More recent treaty provisions of this character tend to impose an apparently less strict liability and to contain more detailed rules. An example is the Indus Waters Treaty 1960 [85] between India and Pakistan. Article IV (10) reads:

'Each Party declares its intention to prevent, as far as practicable, undue pollution of the waters of the Rivers which might affect adversely uses similar in nature to those to which the waters were put on the Effective Date, and agrees to take all reasonable measures to ensure that, before any sewage or industrial waste is allowed to flow into the Rivers, it will be treated, where necessary, in such manner as not materially to affect those uses: Provided that the criterion of reasonableness shall be the customary practice in similar situations on the Rivers.'

To revert to the Boundary Waters Treaty, its main significance for the present purpose is the International Joint Commission which is created to supervise the implementation of the Treaty as a whole, including the prohibition of pollution. The Commission consists of representatives of both countries. Its powers with respect to water pollution are mainly of a technical, investigatory kind. Although it could not be said that its efforts have prevented or ended the pollution of all waters within its jurisdiction, its reports on pollution have at least served to identify the problems that exist.

The value of international bodies competent to study river pollution and to make proposals for its control is now well recognised in Europe. A number of such bodies have been established by treaties in the last twenty-five years. Thus there are now International Commissions for the Moselle, the Rhine, the Saar, Lake Constance and Lake Geneva, for certain rivers of common concern to Belgium, France and Luxembourg, and for frontier waters between the Federal Republic of Germany and the Netherlands and between Italy and Switzerland [86]. Typical of them is the International Commission for the Protection of the Rhine against Pollution [87]. This is composed of 'delegations of the signatory governments' [88]. The Commission's tasks are to 'prepare' and 'have carried out', as a rule by 'competent national bodies', investigations 'necessary to determine the nature, extent and causes of pollution of the Rhine' and to 'analyse their results'; to 'make proposals to the signatory governments on measures for protecting the Rhine against pollution'; and to 'prepare the guidelines for possible arrangements between the signatory governments concerning protection of the Rhine waters' [89]. Decisions are taken unanimously [90]. The Commission is, in a sense, a standing conference of interested states with the important power of initiating, given common consent, investigations into river pollution and to recommend means for its control. It is, unlike, for example, the Court of the European Communities, not a supra-national body with powers of decision binding upon signatory states and persons within them. Nor has it any 'police' powers. It has no competence to require a signatory government to take action to prevent or reduce pollution and to penalise it if the action is not taken and no direct control over the activities of companies and other private persons. These comments are not intended to suggest that the Commission, which is typical of the others referred to in Europe, is of no real value. Such bodies are without question an important co-operative step in the right direction. Equally clearly, however, they depend for their effectiveness upon the continuing goodwill and mutual self-interest of the countries concerned.

An approach of a somewhat different kind is to be found in the Draft Convention on the Protection of Fresh Water against

Pollution [91] proposed within the Council of Europe in 1969. The Draft Convention covers all 'international drainage basins' within the territories of the countries that would become parties to it and proposes a system of judicial remedies for persons injured by river pollution caused contrary to its terms. The Draft adopts, very largely, the definitions of 'water pollution' and 'international drainage basins' in the Helsinki Rules. It is more specific than the Rules in the sense that it lists the river uses that are safeguarded. These are the production at a reasonable cost of drinking water of good quality; the conservation and development of aquatic resources, including both fauna and flora; the production of water for industrial purposes; irrigation; use by domestic animals and wildlife; and recreational amenities, with due regard for health and aesthetic requirements. Parties to the Convention would, 'wherever possible, agree to establish and maintain standards of quality for the waters of an international drainage basin extending over their territories' and, 'where appropriate in the circumstances, establish joint commissions to regulate usage of such waters'. As to remedies, the Draft provides for the reference of a dispute between countries to 'the appropriate joint commission' or other agreed body or, if this proves impossible, to an ad hoc arbitral tribunal 'at the application of any one or more of the Contracting States concerned'. Additionally, '[a]ny person who suffers damage in any contracting state arising from water pollution in any other contracting state shall be entitled to compensation . . .' The procedure proposed for obtaining such compensation is a claim brought by the injured person 'in the courts of the responsible state'. Each party to the Convention would undertake that its courts had jurisdiction to hear such claims. The Draft Convention, as thus proposed, is ambitious both in its plan to cover more than one 'international drainage basin' and in its provision for compulsory arbitration between parties to a dispute and for remedies for private persons. Perhaps not surprisingly, the Council of Ministers of the Council of Europe thought that it 'did not provide a valid basis for an international instrument' [92] and has asked for it to be revised.

Development of the Law

If the question were put to an international tribunal at the present time it would probably hold that customary international law does contain a rule of a general character requiring one riparian country to have reasonable regard for the interests of another in its use, or tolerance of use, of international rivers flowing through its territory and that this requirement relates, *inter alia*, to interests and uses affected by and involving pollution. It is difficult to be more precise at this point of time or to imagine that customary international law can take the matter much further. As in the case of air pollution,* the future probably lies in treaty regimes that not only restrict pollution but also provide effective machinery for the enforcement of the restrictions they impose. Whether this machinery should include a body independent of governments with powers of investigation of its own and binding enforcement powers is not certain. It may be that concern about the environmental effects of river pollution has made such an impression upon governments that, in respect of some rivers at least, advisory technical commissions of the sort described are sufficient. Otherwise, some independent international institution with jurisdiction to act on its own initiative or upon complaints from other parties or from individuals who have allegedly been injured is called for. To the large extent that problems are peculiar to particular rivers, such machinery might be best limited to one international drainage basin, applying standards that are reasonable in the particular circumstances. This leaves open, of course, the question of the pollution of national rivers. This could be dealt with on a human rights basis, by means of a new treaty or a protocol to an existing treaty such as the European Convention on Human Rights. Again, effective enforcement procedures would be of paramount importance.

* See above, Chapter 4B.

References

A

1. In the sense that no more has to be proved to establish the right than the ownership of the riparian land in question.
2. See, e.g., *Crossley* v. *Lightowler* (1867) LR 2 Ch 478.
3. *Taylor* v. *Bennet* (1836) 7 C & P 329. The discharge of completely in-

nocuous matter is also probably not actionable: *Kensit* v. *G. E. Rlwy Co.* (1884) 27 Ch D 122.
4. Under the Prescription Act, 1832, or possibly by prescription at common law.
5. *Young* v. *Banker Distillery Co.* [1893] AC 673, at p. 678.
6. Cf. in relation to atmospheric pollution, the case of *Att-Gen.* v. *PYA Quarries Ltd* [1967] 2 QB 169.
7. *Jus tertii* can always be pleaded as a defence in this kind of proceeding: *Stockport Waterworks Co.* v. *Potter* (1864) 3 H & C 300; *Paine & Co.* v. *St Neots Gas Co.* [1939] 3 All ER 812.
8. *Ballard* v. *Tomlinson* (1885) 29 Ch D 115, at p. 121.
9. *Hubbs* v. *Prince Edward County* [1957] 8 DLR 2nd, 394 (Canada).
10. *Dell* v. *Chesham UDC* [1921] 3 KB 427.
11. *West* v. *Bristol Tramways Co.* [1908] 2 KB 14.
12. *Harrington* v. *Derby Corporation* [1905] 1 Ch 205.
13. *McIntyre* v. *McGavin* [1893] AC 268.
14. [1953] 1 All ER 179.
15. A property right which is protected by the law as are other rights of ownership.
16. *Rugby Water Board* v. *Walters* [1966] 3 All ER 497.
17. Public Health Act, 1875; Rivers Pollution Prevention Act, 1876.
18. Originally established as 'river boards', by the River Boards Act, 1948, and reorganised by the Water Resources Act, 1963.
19. 'Stream' is defined for this purpose to include any river, stream or water-course, whether natural or artificial, but it does not include a public sewer or a lake or pond which does not discharge into a stream, and it also does not normally include tidal waters; 1951 Act, s. 11(1). Accidentally allowing polluting matter to enter a stream does not amount to an offence under this section: *Impress (Worcester) Ltd* v. *Rees* [1971] 2 All ER 357. However, deliberate acts of disfiguring apparatus (such as settlement tanks) which, if not properly maintained, would result in the escape of polluting matter into a stream, would be caught by the section: see the decision of the House of Lords in *Alphacell Ltd* v. *Woodward* [1972] 2 All ER 475.
20. Except in the case of a local authority discharging by a means of a sewer; it is not open for them to argue that they were not aware that the matter discharged was noxious or polluting: 1951 Act, s. 2(1).
21. 1951 Act, s. 7(1).
22. Ibid.
23. 1961 Act, s. 1.
24. 1951 Act, s. 7(7).
25. Public Health Act, 1936, s. 20.
26. I.e., the county borough or county district council.
27. Public Health Act, 1936, s. 34; subject to no noxious matter being caused to flow into a sewer: s. 27, ibid.
28. Ibid., s. 23.
29. Ibid., s. 14.
30. Public Health (Drainage of Trade Premises) Act, 1937, s. 14(2), and Public Health Act, 1961, s. 4.
31. 1937 Act, s. 2.
32. Public Health Act, 1961, s. 60.
33. See definition in s. 135(1) of the Water Resources Act, 1963.

34. Water Resources Act, 1963, ss. 72–5.
35. There are no less than ninety-five areas specified in this schedule.
36. Where obnoxious matter enters waters accidentally, this does not amount to 'causing' to flow: *Moses* v. *Midland Rlwy* (1915) 79 JP 367; nor is the owner of an oil depot liable if a valve is opened by a stranger: *Impress (Worcester) Ltd* v. *Rees, The Times*, 19 Feb. 1971.
37. Salmon and Freshwater Fisheries Act, 1923, s. 8.
38. Ibid., s. 9.
39. Cemeteries Clauses Act, 1847, s. 20.
40. Public Health Act, 1875, s. 68.
41. See Third Schedule, para. 42.
42. Water Act, 1945, Third Schedule, para. 71.
43. See Chapter 4 dealing with the prevention of air pollution, at p. 71.
44. Alkali, etc., Works Regulation Act, 1906, s. 4.
45. Diseases of Animals Act, 1950, s. 78.
46. Water Act, 1945, s. 19.
47. Ibid., s. 18.
48. Ibid., s. 22.
49. 'Radio-active material' is defined in s. 18 of the Act of 1960.
50. Of Housing and Local Government; now the Secretary of State for the Environment.
51. In these cases an authorisation is required for disposal only; not for accumulation.
52. Radio-active Substances Act, 1960, s. 9(1).
53. Ibid., s. 9(3).
54. *Supra*, p. 114.
55. As, for example, where pesticides from riparian farm land are washed off by rain water into the river, or an accidental spillage of petrol or diesel oil (perhaps from an overturned vehicle as a consequence of a road accident) escapes into a ditch leading to the river.
56. As amended by the Protection of Animals (Amendment) Act, 1927, s. 1, and the Protection of Birds Act, 1954.
57. Cf. observations at p. 114 above.
58. Destructive Insects Act, 1877; Destructive Insects and Pests Acts, 1907 and 1927; Prevention of Damage by Pests Act, 1949.
59. Poisons Rules 1952 (SI 1952, No. 2086) rules 2(1) and 14(2)(b).
60. Prevention of Damage by Rabbits Act, 1939.
61. See, for example, 'Further Review of Certain Persistent Organochlorine Pesticides used in Great Britain', published by HMSO.
62. Resolution (70)24, passed in 1970.
63. Only some of the members of a water authority will be nominated by local authorities in the area; *all* the members will be appointed by the Secretary of State.
64. The river 'basin' authorities also would have continued in existence, but all authorities would (presumably) have been liable to receive directions from the national authority. Under this proposal the 'basin' authorities would have had additional powers including river and aquifer management, flood prevention and the conveyance and treatment of sewage in bulk.
65. 'Escape' of pesticides to the plaintiff's land from the land of the defendant would be actionable under the Rule in *Ryland* v. *Fletcher* (1866), without any proof of negligence.

B

66. For the most part, the term 'river' is used in this section for convenience. The comments that are made apply equally to other kinds of natural inland waters.
67. See Moore, 1 *Digest of International Law* 653.
68. See Dealey, 23 *AJIL* (1929) 307.
69. Whiteman, 3 *Digest of International Law* 943.
70. See the Report of the Committee on the Uses of the Waters of International Rivers of the International Law Association, *Report of the 52nd Conference of the International Law Association*, Helsinki (1966), p. 477, at p. 487.
71. 24 *ILR* 101.
72. Ibid., p. 123.
73. 9 *Annual Digest* 120.
74. Ibid., p. 121.
75. *International Law*, Vol. 1, 8th Edition (1955), para 178c. Enforcement of such a limitation would have to be by negotiation, arbitration, etc. For an example of the contrary view, expressed by an American writer, see Briggs, *The Law of Nations*, 2nd Edn (Stevens, 1952), p. 274.
76. 24 *Annuaire* (1911), 365.
77. See the *Report of the 52nd Conference of the International Law Association*, Helsinki (1966), p. 477 *et seq*.
78. Article II, Rules.
79. Op. cit. at n. 77, above, p. 500. The same meaning can probably be given to 'substantial damage' in Article X(1)(b).
80. Article IV, Rules.
81. Op. cit. at n. 77, above, p. 487.
82. Ibid., pp. 500–1.
83. Article XI, Rules.
84. Op. cit., in n. 77, above.
85. 55 *AJIL* (1961), 797.
86. See *Fresh Water Pollution Control in Europe*, Council of Europe (1966), p. 98 *et seq*.
87. For the English text of the 1963 Treaty between the FRG, France, Luxembourg, the Netherlands and Switzerland placing it on a formal basis (it had existed in fact since 1950), see ibid., Appendix 6. The Treaty came into force in 1965.
88. Article 3, 1963 Treaty.
89. Article 2, 1963 Treaty.
90. Article 6, 1963 Treaty.
91. CE Doc. 2561, 12 May 1969.
92. CE Doc. 2904, p. 6.

Chapter 7

Land Dereliction: Technical and Administrative Aspects

by BRIAN HACKETT

Professor of Landscape Architecture, University of Newcastle-upon-Tyne

Types of dereliction

Land dereliction is interpreted fairly widely by the general public, ranging from an area of land or an object with an appearance of decay to an area of land unable to support the natural vegetation and wildlife species of the locality. There is, however, an element of duty or responsibility inherent in dereliction and its avoidance; current thinking about the environment places a duty on developers of land to take measures to avoid future dereliction, such as at the manufacturing stage, or through staged remedial works immediately after dereliction occurs, and will accept nothing less than restoration to a state of fertility.

There are many kinds of derelict land which have been left behind after the persons or industry responsible have ceased to operate, and each kind has its own special problems. Various classification systems have been evolved, such as those by S. H. Beaver,* P. W. Bush,† and J. R. Oxenham‡ but the following comprise the most frequently met types with some of the special problems associated with their reclamation.

Coal mines. The components of a derelict coal mine usually comprise a waste heap, scattered buildings and plant in bad repair, and large areas of land once used for stacking coal, railway sidings,

* S. H. Beaver, 'Land reclamation after surface mineral working', *Town Planning Institute Journal*, Vol. XLI, No. 6 (1955) and others.
† P. W. Bush, 'Spoiled lands to the south-east of Leeds', Proceedings of the Derelict Land Symposium, Iliffe Science and Technology Publication Ltd (1969).
‡ J. R. Oxenham, *Reclaiming Derelict Land* (Faber, 1966).

etc. The waste heap may obstruct light and the view from adjoining houses; it may be subject to combustion and produce obnoxious fumes; the material of which it is composed may be toxic to the extent at which vegetation is unable to survive. On the other hand, a well-formed heap slowly acquiring a vegetation cover through natural processes and without the problems mentioned, is not in itself necessarily an ugly or offensive element, especially if it is set in a dull landscape. There is, in fact, some truth in the statement that the real cause for complaint lies in the untidiness and dereliction around the heap, rather than in the often clearly defined form of the heap. Although subsidence problems from underground workings are usually associated with the roads and farmland around the pit head, these can occur within the surface working area and are likely to interfere with the natural drainage arrangements.

Various administrative problems are met in connection with coal mines, such as a requirement that the land should be restored to its original state on the termination of a lease arranged possibly ninety-nine years ago, and the existence of public utility services with wayleaves across the site.

Current practice in reclaiming waste heaps favours grading to produce even and gentle slopes which can later support agriculture, industry or housing. On some sites, the bulk of the heap is too large to be distributed satisfactorily over the surface area of the mine, with the result that adjoining land must be acquired to enable a satisfactory scheme to be designed. Problems also arise during grading works, particularly when zones of combustion are met or when the material is easily blown about by strong winds thus creating a temporary dust problem for adjoining areas.

The often varying nature of the material in any one waste heap, which is difficult and expensive to pre-determine with exactitude, raises problems in the distribution of the material so that toxic and fused materials are buried below the finished levels.

Iron and copper ore mines. The fact that most of the waste material from ironstone and copper workings is warm in colour, compared with the grey and black of coal workings (excepting the red burnt shale), makes them on the whole less repellent in appearance. The debris from copper working can be spectacular in appearance, but this questionable advantage is offset by the toxicity level which is often inhospitable to vegetation.

Until the inception of the Ironstone Restoration Fund (Mineral Workings Act, 1951), the opencast method of extraction often left behind a 'hill and dale' landscape of parallel ridges and valleys; unless expensive regrading of each ridge into the valleys was carried out, the usual method of restoration was tree planting on the uneven land. Today, with the experience of the Opencast Executive of the National Coal Board, it is possible to carry out this form of mining in such a way that restoration to something close to the original landscape follows quickly upon extraction.

The problem of toxicity mentioned previously often occurs in pockets rather than over the entire derelict area, and a first stage solution is to concentrate upon the areas most likely to respond to cultivation, planting and seeding, weaving them into a pattern around the toxic areas and leaving the latter to respond in the course of time to natural processes.

Clay and gravel workings. Compared with the coal and metalliferous ore extraction industries, clay and gravel workings leave large holes in the ground with only a small amount of overburden to fill back to somewhere near the original level. Thus, unless the working can be used as a fill area for industrial waste from elsewhere, the main problem is a hole in the ground. The establishment of vegetation is usually much less of a problem than with coal or metalliferous ore dereliction.

In the case of a 'wet' pit, which occurs when the bottom of the pit is an impervious material and there is no drainage outlet, there is a sterilisation of the land as a producing medium unless the pit can be filled, but its recreation potential can be high. In the case of a 'dry' pit, the stripping of the topsoil just before extraction of the clay or gravel takes place and its speedy deposition on the floor of the pit is an additional task rather than a problem.

The noise from the screening plant of gravel workings is a problem when people are living or working close by, especially as the plant tends to be located near one of the boundaries.

Quarries. The problem of the after-appearance of a stone quarry is simplified when the design of the working operations is in sympathy with a landscape solution. For example, a cut into a hillside should take place through a narrow opening with the quarry widening out behind, and the shape of the quarry should relate to the surrounding topography. Also, the situation in some quarries is such that overburden has to be removed to get at the stone and this can be deposited in such a way that the quarry is screened from nearby and long-distance views. Both in the examples of a new quarry subject to landscape conditions in the planning permission and of a worked-out quarry, tree planting will in time prove a valuable ameliorative measure.

Quarries often constitute potential danger situations, such as vertical drops, falling material and deep water. These situations do not remain if the quarry can be filled with spoil; if domestic refuse is used for filling, problems often arise from lack of drainage and the difficulty of sealing the refuse after each deposition when the 'water table' is high, and there is sometimes a fire risk problem.

Subsidence. Agricultural land subject to subsidence from deep mining operations often has an interesting appearance from the ponds and areas of marsh vegetation which result. Nevertheless, from the agricultural point of view, such areas are derelict because of the interference with the natural drainage system and the sharp undulations of the ground which make cultivation operations difficult. It is possible to fill or grade the hollows back to the original level, but there is often the possibility of further subsidence taking place. The interference with the natural drainage also makes subsidence areas unattractive for economic forestry.

Industry. The kinds of dereliction left behind by abandoned factory industrial sites are extremely varied. Apart from the poor appearance, the major problem is often the attraction of such areas for vandalism and to children who innocently use them as play areas. The buildings, plant and machinery may have their own special problems, but these can usually be overcome by demolition methods. Sometimes toxic areas are left and create a special problem. It is also difficult to obtain information on the runs of underground services and pipelines, especially if the site has been abandoned for a long time.

Slum clearance. The major problems of slum clearance areas are usually of a planning and administrative nature rather than of a physical kind—rehousing of the displaced persons often involving a difficult operation with social and phasing problems. There is also a vandalism problem when the area becomes derelict.

Compared with many other examples of derelict land, slum clearance areas rarely have the advantage of an open landscape around, which can reduce the visual impact of the dereliction and simplify the design of the reclamation proposals.

Canals and railways. Disused canal and railway systems are rarely such a visual problem as the other types of dereliction, except where associated buildings and plant are allowed to decay. Vegetation on each side of the canal or line soon develops naturally with, however, an associated maintenance problem. The latter extends to the maintenance of structures—if, for example, a canal bank is not maintained, it may burst and flood adjoining land. Also, bridges, both over and under railways, need periodic attention.

The most satisfactory solution to the problem of derelict canals and railways is to open them up for re-use. Failing this solution, conversion into roads or footpaths will justify expenditure on restoration matters.

Beaches and river banks.   Dirty or debris-covered beaches and eroding situations are the two main groups within which dereliction problems lie. A difficulty in making proposals to solve these problems is often one of the lack of a financial return from the reclamation work, such as comes from the restoration of derelict areas to agriculture.

Control of Dereliction
The Civic Trust has estimated* that, despite planning legislation and active reclamation projects, the area of dereliction in Britain was increasing at a rate of some 3,500 acres a year. Thus, greater activity in administering the control measures, and in formulating new measures, will be necessary to bring the situation to a state where the total area of dereliction is being systematically reduced. These measures have to cover the dereliction problem so that both the organisations who create dereliction at the place of operation, and those who do so elsewhere by indirect means, are kept in check.

Dereliction in situ.   The advanced planning legislation now operating in Britain enables restoration processes to be enforced quickly by including controls on the use of a site for which a planning application has been made for development. Nevertheless, the success of the legislation largely depends upon the tenacity of a planning authority to insist upon the controls and to ensure they are instituted if the central government department concerned with planning will uphold the conditions upon which permission has been granted.

The application of controls on the use of land by industry may

* *Derelict Land*, Civic Trust (1964).

well involve additional expense and trouble, although a valid argument is that the industry should benefit in the long run.

The National Coal Board is an example of an industry which takes greater care than its predecessors in reducing the likelihood of future dereliction, and this is being achieved without the special conditions to which other mineral extractors are subject. Also, the Central Electricity Generating Board is very conscious of the dereliction problem caused by the deposition of its waste material —pulverised fuel ash—and makes strenuous efforts to avoid land becoming 'permanently' derelict. Nevertheless, success in the control of dereliction lies in the willingness of industry to adopt a responsible attitude towards the environment and in the determination of the planning authorities to insist upon the application of the necessary conditions. In this respect, the planning officers must have the backing of the planning committee and the central government department concerned. There is also the planning authority's problem of insufficient staff to maintain regular checks on whether the conditions are being observed.

In certain circumstances, new techniques of operation on the part of industry may be necessary to avoid the creation of a further dereliction problem; these techniques may cost money on a narrow assessment, and are likely to be resisted by the industries concerned. An example of co-operation in this respect is the shaping and grading of coal mine waste heaps by controlled tipping so that the new contours do not inhibit several possible future uses. Some measures to minimise the effects of quarrying upon the environment were referred to earlier.

The rapid technological and economic changes which now take place often mean the closing down of an industrial site with no successor to keep the buildings and plant in good repair. This situation extends to some mining villages which were located in remote places in so far as industry and population are concerned. Because of this latter specialised locational factor, planning authorities are unable to produce a realistic adjustment of industrial location to fill the employment gap; the solution inevitably has to be the re-location of the population and the removal of the houses and roads which are likely to be in a reasonable state of repair.

Indirect dereliction. Dereliction at a distance from the cause or source is often due to some form of pollution, and it is not surprising that the chemical industries are often the responsible parties, or, if not, are particularly vulnerable to charges made against them. The pollution is usually water- or airborne, and can kill life in rivers and streams, and erode away building materials. The source is often difficult to isolate, especially when the damage occurs at a considerable distance away as, for example, the atmospheric pollution from petrol- and diesel-powered engines which may come from an oil refinery some hundreds of miles away; there is, also, the pollution from the refinery itself.

Some forms of pollution can be beneficial if subjected to treatment; sewage and waste disposal properly treated and deposited can help to build up fertile soils and, in so doing, conform with the natural cycle of events.

Pollution from sewage illustrates one of the major problems in controlling indirect dereliction. In most cases, the pollution has continued and increased over a long period of time so that its removal is likely to be a very expensive and difficult matter. Large cities, located on rivers, which discharge untreated sewage into the rivers now find that many millions of pounds will be needed, and much disruption of the built environment will take place, before the indirect dereliction of the river, its banks and very probably adjoining coastal beaches is removed.

The control of indirect dereliction is complicated by the number and variety of legal provisions, as well as by the remoteness of the source. Public health, river and stream pollution and clean air are among many matters that fall within the law, and the planning legislation can be used to control future dereliction from new remote sources. Even when legislation is available, it has often been ineffective because of the threat to local employment if restrictions are put upon the operation of an obnoxious process.

At the international level, the problem of pollution resulting from the actions of another country and indirect dereliction are now fortunately beginning to command attention.

The Loss to the Nation

The Civic Trust's estimate of the losing battle over the extent of dereliction was a measure of the land lost annually which might otherwise have been used for some useful purpose like industry or housing. There are also less tangible, but important, effects of dereliction as a loss to the nation.

Health. Little attention was paid in the past to the effect of derelict land upon health, possibly due to the more spectacular health hazards at the source, such as silicosis as a result of coal dust in the coal mine compared with the unpleasantness of a burning spoil heap which would depress and irritate the nearby inhabitants. Also, land left derelict may produce polluted run-off into streams—this may affect both life in the streams, and possibly the local water supply.

The particularly difficult environmental matters to measure in areas of dereliction are those producing small, but long-term, physical and psychological effects such as those resulting from the diminution of light to people living near to waste heaps. One of the most striking observations made by people living near to the site of a recently regraded waste heap was, for the first time in their years of occupation, sunshine had entered the house in winter.

Amenity. There is general agreement that the appearance of most derelict areas detracts from the amenity of a locality—so much so that one reason behind the policy of encouraging reclamation was to make such localities more congenial to live and work in, and thus, more attractive to new industry. This statement, however, needs some qualification because there are some derelict sites where vegetation coming in naturally has produced a type of landscape which is easy to walk through, although in a wild state, and very pleasant to observe. Claims have also been made that in certain circumstances a distinctively shaped waste heap has considerable amenity value, for example, a geometrically perfect cone is regarded as a favourable topographical feature in a flat landscape.

Such effects as obnoxious fumes and dust blown from derelict areas, whose health hazard effect may be difficult to prove, certainly result in a loss of amenity by virtue of the discomfort and nuisance value.

Potential resource. The conventional cost–benefit study based upon a quick return of investment rarely applies to the reclamation of derelict land, except in congested urban areas where land values are high, or where a reclaimed site in a rural area happens to be favourably located as regards some new industrial development.

The apparently valid argument that derelict land restored to agriculture will make up for agricultural land taken for urban expansion and motorways is often discounted by the argument that new techniques can lead to increased production from a smaller area of agricultural land. Also, the first argument is obscured by the availability of food from overseas. An argument based on sounder premises is that the reclamation of derelict land increases the nation's resources of both fertile land for future intensive cultivation if the need arises and infertile land in locations which often fit into the programme for urban and industrial development.

There is an increasing recognition of the importance of a balanced wildlife population to a nation's resources, and it must be admitted that some derelict areas which have a partial vegetation cover through natural regeneration provide suitable conditions for wildlife. When such areas are reclaimed it is important to make specific provision for the wildlife population to continue. Also, when derelict areas, unattractive to wildlife, are reclaimed so that soil fertility is built up, they will have a more lasting resource benefit if similar provision is made.

In evaluating reclaimed land as a resource, an allowance should be made for the fact that its full potential will not, in many cases, be fully realised until a period of years has elapsed. In some ways, a newly reclaimed site is in the same position as the landscape at the end of the Ice Age when a considerable period of time passed before a natural drainage pattern and a soil profile had evolved.

Reclamation Techniques

Each new reclamation project is likely to expose new problems and require new techniques to overcome them. Even so, the fundamental basis of successful schemes is that they have been studied as landscape design problems involving provisions for the functioning of healthy landscapes, such as the initial landscape survey and analysis to expose these provisions, and arranging the design to provide for stable surface drainage, soil renewal, topographical forms relating both to the locality and to likely land uses, an adequate cover of vegetation, and planting that will eventually lead to physical and visual improvement of the environment. It should be made clear that separate engineering or agricultural solutions to reclamation problems are unlikely to produce a result which fits into the landscape pattern of a locality.

Vegetation cover. The technique of reclamation most frequently used in the period from the late eighteenth century to the 1950s was aimed at producing a vegetation cover over the derelict site without any grading works. The aim being to improve the appearance at minimum cost, and in the hope that some return might come from the timber or the grazing of sheep and cattle. In most examples, planting or sowing took place directly into the waste heaps or 'hill and dale' areas, whilst a partial grading scheme was occasionally undertaken to produce a level area on top of a waste heap or by smoothing out 'hill and dale' areas, with a topsoil cover brought in.

In general, it can be said that the places where vegetation is coming in naturally are likely to be the most certain places in which to establish new planting. Also, that minor terracing works which could hold back the run-off water are likely to help vegetation to become established. The importance of species trials is stressed, but knowledge is now becoming available which indicates those species which are most likely to be successful over a great many sites. For example, Whitebeam (*Sorbius aria*), Alder (*Alnus glutinosa, A. incana, A. cordata, A. rubra*), White Willow (*Salix alba*), Poplar (*Populus alba, P. canescens*) and Bird Cherry (*Prunus padus*) have been successful on many sites.

Grass establishment on regraded derelict sites has often produced good results after the proper investigations have been made, but it has not been so successful as a cover for ungraded waste heaps. The dry and warm conditions that can develop on the slopes of waste heaps in sunny periods and dry weather are particularly difficult for the delicate initial root systems of grass seeds as they germinate. Successful grass establishment certainly depends upon a proper topographical design which ensures moisture in reasonable quantities over long periods in the top layers of the soil, also upon appropriate cultivation methods, fertiliser treatments and seed mixtures, all of which are matters for expert advice on each site. Nevertheless, certain species and strains figure again and again in successful projects.

Some useful results can be obtained by a blanket planting with a mixture of seedling tree species, and leaving matters to the natural conditions to eliminate those species which find the site uncongenial. This technique overcomes the waiting period while trial plantings produce their answers. There is, of course, an element of wastage in this technique.

The argument has been put forward that planting an undisturbed waste heap will sometimes render it a more prominent object in the landscape than if it was left to weather and acquire a 'patina' of vegetation by natural processes.

Restoration to the original state. Some derelict sites may be subject to the terms of a lease which require that the site should be handed back to the lessor at the end of the lease in its original state, ready for the original use to operate upon it. This could prove to be an expensive operation if full restoration was demanded within a short period of time. Also, circumstances may have changed to such a degree that the original land use would be inappropriate. The usual procedure in such cases is to endeavour to arrive at a negotiated settlement, bearing in mind that the original value of the land may be greatly enhanced as a result of the current local demand to develop more profitable uses than the original use.

The problem is not, however, difficult with sites which have a thin stratum of coal or other profitable mineral, the stratum being removed and marketed and the overburden replaced. Restoration in these cases, apart from the slow growth of trees and hedgerows, follows quickly to a state of fertility which in turn will soon match the original state.

Restoration to a new physical state. In those areas of West Germany where 'brown coal' is extracted by open-cast or strip mining operations, the strata are much thicker than with the thin black coal seams in Britain; thus, restoration must take place at lower levels than those existing before extraction commenced. Also, the scale of the 'brown coal' operations has led to the removal of complete villages and the relocation of the inhabitants elsewhere. As a result, it is possible to design a new landscape for which there are two opposing policies—on the one hand, that of the land user whose aim is to achieve a pattern exactly modelled to his current economic and technological possibilities, on the other hand, that of landscape planning in which the aim is to achieve a long-term healthy and self-maintaining landscape founded upon ecological principles.

A similar opportunity to design a new physical state of the landscape occurs with many of the derelict areas left from the surface operations of deep mining and from deep gravel pits. The opportunities presented by these situations produce results as wide ranging as recreation lakes and 'permanent' agricultural grassland.

The wisest policy is undoubtedly that postulated under the discipline of landscape planning, the aim being to produce a basic *landscape*, rather than an *area of land* capable of being worked upon for some artificial process. In addition, such a basic landscape can provide for the planning authority's demand for flexibility in land-use planning so that a number of different uses could be introduced as the planning situation changes. The provisions made in this landscape planning approach will create shelter, arrange suitable aspects, and include surface drainage considerations. Because of the likely requirement of future land uses, the topography will have the steep gradients limited to small areas. Even when the new use of the site is clearly determined, there is a moral obligation upon landscape designers to plan landscapes which are inherently fertile, conform to natural principles such as the balance between surface drainage and erosion control, and respect the surrounding landscape.

Initiation of a natural recovery process. This technique can take two forms. First, to supplement areas where vegetation has already developed naturally by additional planting, soil improvement, and minor grading works to increase the available moisture to these areas. This particular form of the technique is likely to be suitable for derelict areas which have been undisturbed for a considerable time, and are not likely to be in demand for new uses.

The second form of 'natural recovery' technique would include such provisions as the following:

(a) Grading works designed to direct surface water to the most favourable areas for establishing dispersed concentrations of vegetation from which natural regeneration will cover the intervening areas.

(b) Designing a phased plan for vegetation development based upon the sequence of the succession of plant communities from the bare soils. For example, a clover/grass mixture can prove beneficial to soil improvement, to the retention of soil moisture, and to the provision of shelter for encouraging growth of a seedling transplant—this is a simulation of the natural succession process, but omitting the shrub layer.

(c) Speeding the development of a soil micro fauna population through research into the favourable conditions.

(d) Introducing animals under controls so that restricted grazing and natural manuring encourage the growth of vegetation.

This technique of the natural recovery process is probably the most important field into which research is needed, and is not unlike the lengthy evolution of a richly vegetated landscape from the bare clay or other overburden left after the Ice Age.

Preliminary investigations. These are likely to be numerous, and it is always advisable to build-up a check list on the following lines:

Legal: Statutory: Town and Country Planning, including road, restrictions and proposals. Statutory undertakers' easements across the site. Drainage obligations and possibilities.

Ownership: Boundaries and terms of existing leases.

Public: Rights of way, rights of light.

Services: Location, depths, etc. of gas, water, electricity, sewer, telephone, radio and TV pipes or wires. Transformer and pumping stations, etc.

Topography: Usually necessary to have a contoured site plan prepared from air photographs, and must include an adequate area around the site.

Land: Surface and solid geology. Buildings and plant, underground workings, tunnels, filled areas. Types of material at surface and at depth. Potential for spontaneous combustion. Fertility and erosion status.

Drainage: Levels, capacity and location of possible outfall positions.

Vegetation: Areas and types of natural vegetation. Species flourishing in the locality under similar circumstances.

Design process. The design process will often be influenced by the policy of the sponsoring body or bodies and by detailed requirements in the brief. Irrespective of these influences, special circumstances may require some variation from the customary stages of the design process. Where re-grading is likely to be involved, an early stage will be to work out roughly some alternatives to ascertain the possibilities of cut and fill operations; the most favoured regrading proposal would then be refined to take into account the surrounding topography, the slope/erosion relationship of the different surface materials, levels which cannot be altered such as the minimum cover over water mains, and future land uses where these are known.

Upon the preparation of the preliminary grading design, it will be necessary to test it against the demands of vegetation in terms of aspect, moisture provision, and the distribution of the various 'soil' materials. The surface drainage provisions are likely to comprise sheet run-off from gently graded surfaces, collecting ditches, terraces, flumes, diagonal stone drains on steep slopes, and temporary open drains to prevent the formation of gullies prior to the establishment of a cover of vegetation.

Boundaries to field and woodland units are likely to be decided in relation to the new topography which in turn will indicate the need for shelter belts and the suitable uses for each unit. The sizes of the units should relate to their future maintenance through agriculture, forestry or amenity practices as the case may be.

The matters so far mentioned establish the landscape pattern of the design, and further design work will settle road and footpath access, the location of buildings, watering positions, etc.

Finally, decisions have to be made on the cultivation, fertilising, seeding and planting techniques and upon the species of tree, shrub and groundcover plants to be used. These particular matters are usually determined provisionally at the design stage, and are finalised during the course of the reclamation works upon the results of laboratory tests and trial plantings.

Maintenance. The plans for maintaining reclaimed landscapes can be divided into two classes—those for schemes embodying an immediate and continuing use of the land, and those for schemes with no specific immediate use. With the former, the maintenance will usually comprise agriculture, forestry or a conventional open space maintenance programme for projects like new industries, playing fields and housing. With the latter, a solution may be found through designs which arrange for limited public access controlling vegetation growth in some areas, and for the basis of a self-maintaining complex of vegetation in other areas—the 'soil' conditions might be such or be arranged to accept something akin to the heathlands and commons which are not favourable to the growth of a rich flora, but are fairly resistant to public access. There is also the possibility of arranging intermittent agricultural use, for example, the removal of a hay crop or occasional grazing by a local farmer.

Schemes having a use introduced quickly require an arrangement with the user which will limit and direct his operations for the first years. This is important, even when it has been found possible to let cattle out to graze in less than one year from seeding an area.

Regular soil tests and examinations of the soil profile and vegetation growth are most necessary in the first years, and may lead to advice on the fertilising programme, the grazing limits, and if and when ploughing should take place and a crop be planted. A satisfactory arrangement is for the land to be leased to a farmer on favourable terms, but on condition he conforms with the maintenance programme; a safeguard would be for the delivery of the fertiliser to be the responsibility of the lessor, whilst the lessee applies the fertiliser to the land. Unless the farmer is equipped to undertake forestry operations, the brashing, thinning and other operations required for shelter belts and woodlands are best let on a contract basis, though some local authorities with large areas of derelict land have set up their own forestry planting and maintenance organisations.

An aspect of maintenance which often presents difficulties is the possibility of substantial changes taking place to the reclaimed landscape after the initial contract is ended. Newly formed water courses are particularly vulnerable to flash floods and can erode severely after they appear to have reached the point of stability. Difficulties of this kind are acknowledged by the government grant authority, and further assistance is likely to be forthcoming, for example, additional works to develop grassed areas can be considered up to three years after the contract.

Costs. Negotiations over the costs of reclamation would be eased if the future and continuing value of the reclaimed land could be accounted for in the figures. For example, a derelict site, producing no income, may cost as much as £100,000 to reclaim, but this should be set against the annual income it can produce for all time after reclamation. The arguments for spending money on reclamation would also be considerably strengthened if a method could be devised to assess environmental improvement in monetary terms.

Reclamation costs are bound to vary as between one project and another. In the north-east the cost of eight projects, adjusted to 1971 values, ranged from £510 acre to £4,800 acre. The major cost of most schemes is in the bulk movement of material, and the design policy will take into account proposals with limited grading works, and probably limited future uses as regards financial return, and evaluate them with major grading works and a subsequent profitable use of the land.

Experience has demonstrated that the securing of tenders through the quantity surveying process is most beneficial. Apart from the closer tendering which is likely to occur, the preliminary estimates on the basis of draft bills of quantities simplify the negotiations with the central government department concerned with grant approval. A bill of quantities is also an invaluable document upon which to base the cost of variations which are prone to arise in reclamation work as hidden items are revealed.

A further cost factor is the timing of a project. Many reclamation schemes are well drained as regards mounds and spoil heaps, and these can be worked upon when conditions for normal grading operations are nearly impossible elsewhere. Thus, mid-winter timing sometimes produces surprisingly low tenders.

In order to qualify for grant, the various works proposed must be necessary reclamation, in other words proposals of a kind normally associated with aesthetics are only likely to be approved for grant if they also have a land use or stability value.

Control of Dereliction

The major kinds of control lie with the powers of the Planning, the Public Health, other Water Resources and the Clean Air Acts. Especially in the case of the Planning Acts, some control can be exercised upon the location of the potential offender, and there are specific powers under the other Acts limiting pollution and which are likely to be acceded to by technological improvements to the manufacturing processes. In addition to controlling the location of an industry, powers under the Planning Acts can restrict the area of land to be used and the direction on which physical expansion of the plant or extraction workings takes place. They can also require new plantations and mounds to be introduced for screening purposes, and an agreed plan of the phasing of expansion and restriction.

In general, it can be said that legislation is available in Britain to bring about the restoration of derelict landscapes and for controlling new damage to the landscape. The slow progress in reclaiming the old and preventing new dereliction is due in part to the shortage of landscape design personnel capable of handling the design and technical problems, and to the lack of drive on the part of some local authorities in embarking upon an adequate number of new projects. It must also be admitted that some planning authorities are not administering the powers available to them with sufficient determination.

References

J. Atkinson, 'Ventilation of mines', *North of England Institute of Mining Engineers. Transactions.* Vol. VII (1859), Andrew Reid, Newcastle upon Tyne, pp. 133–67.

J. W. R. Adams, *The Restoration of Disused Excavations in the Dartford-Northfleet Area and the Disposal of Industrial and Domestic Waste from South-east London and North-west Kent* (Kent County Council, 1954).

John Barr, *Derelict Britain* (Penguin, 1969).

A. M. Bauer, *Simultaneous Excavation and Rehabilitation of Sand and Gravel Sites* (University of Illinois. National Sand and Gravel Association Project No. 1, 1965).

S. H. Beaver, 'Rehabilitation of derelict land', *Nature* (1950), p. 166.

———, 'Land reclamation after surface mineral working', *Journal, Town Planning Institute*, Vol. 41 (1955).

R. B. Beilby, J. W. Blood, J. Casson, W. M. Davies, I. B. Hall, W. J. Rees, R. F. Wood and others, *The Rehabilitation of Areas Biologically Devastated by Human Disturbance*, Report of the sixth technical meeting of the International Union for the Conservation of Nature and Natural Resources (1956).

R. D. Blakely, *Reference List for Reclamation of Strip Mine Areas* (US Dept Agr., 1964).

K. L. Bowden, *A Bibliography of Strip Mine Reclamation 1953–1960* (Univ. Mich. Dep. of Conservation, Ann Arbor, 1961).

J. K. Brierley, 'Some preliminary observations on the ecology of pit heaps', *J. Ecol.*, Vol. 44 (1956), p. 383.

R. A. Briggs, *Landscape Reclamation. Implementation* (Inst. of Adv. Arch. Studies, York, 1966).

British Soil Survey Staff. *Field Handbook*. Soil Survey of Great Britain. Rothamsted, No. 12 (1960).

British Standards Institute. B.S. Code of Practice no. 303. *Surface Water and Subsoil Drainage* (BSI, London, 1952).

K. Browne, 'Dereliction', *Architectural Review*, Vols 118 and 119 (1955/6).

E. F. Button, 'Establishing slope vegetation'. Reprinted from *Public Works Magazine* (1964).

P. Cadbury, 'Land reclamation in the black country', *Town and Country Planning*, Vol. 23 (1954).

Carruthers and Cowan, 'Planning the restoration of an open cast coal site', paper to the *Operational Research Society, Annual Conference* (1966).

G. Christian, *Tomorrow's Countryside* (Murray, 1966).

Civic Trust, *Derelict Land; a study of industrial dereliction and how it may be redeemed* (London, 1964).

W. G. Collins and P. W. Bush, 'The definition and classification of derelict land', *Journal, Town Planning Institute*, Vol. 55 (1969), No. 3.

Commonwealth Bureau of Soils, Bibliography on spoil bank soils and their reclamation (1965–1966), No. 1027CBS, 20 (1965).

G. W. Cooke, *The Control of Soil Fertility* (Crosby Lockwood, 1967), pp. 125–9.

T. Cooper, *Practical Land Drainage* (Leonard Hill, 1965).

The Countryside in 1970 Study Group 12, *Reclamation and Clearance of Derelict Land*, Royal Society of Arts, The Nature Conservancy, The Council of Nature (Garden City Press, 1970).

R. J. Cowan, 'Ironstone workings and land restoration', *Chartered Surveyor*, Vol. 94 (1961).

G. Cowley, *Bedfordshire Brickfield: a planning appraisal* (Bedfordshire County Council, 1967).

J. B. Cullingworth, *Town and Country Planning in England and Wales* (Allen & Unwin, 1967).

W. M. Davies, 'Land restoration following mineral extraction and deposition of waste materials', *Journal, Royal Agricultural Society*, Vol. 129 (1961).

DSIR, *Water Pollution Research* (HMSO, 1939 *et seq.*).

M. F. Downing, 'Reclamation of derelict landscape', *Planning Outlook*, new series Vol. 3 (1967), pp. 38–52.

Durham County Council, *Derelict Land in the North East* (1965).

N. Fairbrother, *New Lives New Landscapes* (Architectural Press, 1970).

R. M. Finch, 'Waste materials and land reclamation', *Contractors' Record*, Vol. 68 (1957).

A. H. Fitton, J. Gibbons and others, 'Experiments in the rehabilitation of open-cast coal sites', *Experimental Husbandry*, Vol. 4 (1959).

F. Flintoff, 'Land reclamation; the use of earth-moving machinery for house refuse', *Public Cleansing and Salvage*, Vol. 44 (1954).

Forestry Commission, *Industrial Waste Land and its Afforestation and Reclamation*. A bibliography of British references (1963).

D. T. Funk, *A Revised Bibliography of Strip Mine Reclamation* (US Dept. Agric. For. Service, Central States Forest, Experimental Station, Columbus, Ohio, 1962).

J. F. Furness, 'Tipping in wet pits', *Municipal Journal*, 6 August (1954).

T. R. G. Gray and S. T. Williams, *Soil Microorganisms* (Oliver and Boyd, 1971).

G. T. Goodman (ed.), 'The 1967 sub-committee survey of the nature of the technical advice required when treating land affected by industry', *Journal of Ecology*, Vol. 55.

G. T. Goodman, R. Edwards and J. M. Lambert, *Ecology and the Industrial Society* (Blackwell Scientific Publications, 1965).

N. Goodland, 'Gravel pits into woodlands', *Town and Country Planning*, Vol. 26 (1958).

B. Hackett, 'Basic design in land form', *Journal of ILA* No. 49 (1960), pp. 7–9.

——, Land modelling. *Public Works and Municipal Service Conference* (1964).

——, 'Landform design and cost factors', *Landscape Architecture*, Vol. 54 (1964), pp. 273–5.

——, 'Earthworks and ground modelling', Institute of Landscape Architects, *Techniques of Landscape Architecture*, by A. Weddle (ed.) (Heinemann 1967), pp. 55–72.

B. Hackett and C. J. Vyle, A report on the alternative schemes for the redemption of Percy Pit, Lemington, in the Urban District of Newburn, incorporating recommendations relating to the redemption of the heap. Unpublished 6 pps.

B. Hackett (ed.), *Landscape Development of Steep Slopes* (Oriel Press 1972).

I. G. Hall, 'Ecology of pit heaps', *J. Ecol.*, Vol. 45 (1957), pp. 689–720.

——, 'The ecology of disused pit heaps in England', *J. Ecol.*, Vol. 54 (1959), p. 689.

T. U. Hartwright, *Planting Trees and Shrubs in Gravel Workings* (Sand and Gravel Association of Great Britain, 1960).

H. E. Hayward and C. H. Wadleigh, 'Plant growth on saline and alkali soils', *Adv. in Agron.*, Vol. 1 (1949), pp. 1–38.

K. J. Hilton (ed.) *The Lower Swansea Valley Project* (Longmans, 1967).

I. V. Hunt and D. P. Farrant, 'Farming an ash dump', *Journal, British Grasslands Society*, Vol. 10 (1955).

Institution of Municipal Engineers, *Mining Subsidence* (IME, 1947).

P. C. G. Isaac (ed.), 'The treatment of trade-waste waters and the prevention of river pollution', *Contractor's Record* (1957).

W. A. Jackson, 'Physiological effects of soil acidity', *Soil Acidity and Liming*. Pearson and Adams (eds), Monograph No. 12 (Am. Soc. Agron., Madison, 1967), pp. 43–124.

C. Johnson, 'Practical operating procedures for progressive rehabilitation of sand and gravel sites', Project No. 2 1964–5 (University of Illinois/National Sand & Gravel Association of America, 1966).

W. Knabe, Contributions to bibliography on reclaiming mined areas (in German, English Summary). Wiss. Z. der Humboldt-Univ. Berlin, *Math. Nat Reihe*, Vol. VII (1957–8), pp. 291–304.

———, 'Methods and results of strip mine reclamation in Germany', *The Ohio Journal of Science* (1964).

W. L. Kubiena, 'Animal activity in soils as a decisive factor in establishment of humus forms', *Soil Zoology Proc. Univ. Nottingham*, Vol. 1 (Butterworth, 1955), pp. 73–83.

W. Kühnelt, *Soil Biology* (Faber and Faber, 1961).

J. R. Lake, *Unburnt Colliery Shale—Its Possible use as Roadfill Material* (Road Research Laboratory MOT, Crowthorne, 1968).

Lancashire County Council, *Derelict Land Problems* (Lancs. CC, 1962).

Ministry of Health: Committee on Restoration of Land Affected by Iron Ore Working: Report (1939).

Ministry of Housing and Local Government, *Bibliography No. 107:* Derelict Land: a select list of references (1957).

———, *Bibliography No. 107:* a select list of references (1963).

———, *Bibliography No. 107, Addendum* (1965).

———, Explanatory memorandum on grants for the reclamation of derelict, neglected or unsightly land under the Industrial Development Act, 1966 and the Local Government Act, 1966 (1967).

———, *Refuse Storage and Collection* (HMSO, 1967).

———, *The Protection of the Environment* (HMSO, 1970).

———, Explanatory Memorandum on the Ironstone Restoration Fund (1955) (Revised).

———, Technical Memorandum No. 3—Mineral Working (1955).

———, Technical Memorandum No. 7—Derelict Land and its Reclamation (1956).

———, Technical Planning Memorandum—Pulverised Fuel Ash (1958).

———, Circular 26/59—Colliery Spoil-heaps (1959).

———, Circular 30/60—Local Employment Act; Rehabilitation of Derelict, Neglected or Unsightly Land (1960).

———, The Control of Mineral Working (1960).

———, Pollution of Water by Tipped Refuse (1961).

———, Technical Committee on the Experimental Disposal of House Refuse in Wet and Dry Pits: Report (1961).

Ministry of Town and Country Planning: A. H. Waters, *The Restoration Problem in the Ironstone Industry in the Midlands* (1946).

———, Advisory Committee on Sand and Gravel: Reports, parts 1 and 18 (1948 and 1953).

———, Country planning, reclamation of mined land. *Bibliography No. 84* (1949).

L. R. Moore, 'Some sediments closely associated with coal seams', *Coal and Coal Bearing Strata* by Murchison and Westoll (eds), (Oliver & Boyd, 1968), pp. 105–23.

F. A. Neustein, Report on Forestry Commission trial plantings on Pumpherston shale bings (1970).

D. E. Newton, 'Clearance of concrete from disused airfields', *Journal, Town Planning Institute*, Vol. 46 (1960).

J. F. Oxenham, 'Land reclamation', *Journal, Town Planning Institute*, Vol. 53, No. 8 (1967), pp. 344–7.

J. R. Oxenham, *Reclaiming Derelict Land* (Faber, 1966).

B. Pickersgill, *Growth of Plants on Derelict Pit Heaps* (Thesis, Univ. Newcastle, 1971).

R. D. Pinchin, Survey of Plantations on Open-cast Ironstone Mining Areas in the Midlands. Reports, Forestry Research, Forestry Commission (1950–2).

W. J. Rees and G. H. Sidrak, 'Plant growth on fly ash', *Nature*, Vol. 176 (1955).

W. J. Rees and A. D. Skelding, 'Grass establishment on power station waste', *Agriculture*, Vol. 59 (1953).

J. A. Richardson, *The Ecology and Physiology of Plants Growing on Colliery Spoil Heaps, Clay Pits and Quarries in County Durham* (Thesis, Univ. Durham, 1956).

——, 'Derelict pit heaps and their vegetation', *Planning Outlook*, Vol. 4 (1957), 15.

——, 'The effect of temperature on the growth of plants on pit heaps', *J. Ecol.*, Vol. 46 (1958), p. 537.

J. A. Richardson and E. F. Greenwood, 'Soil moisture tension in relation to plant colonisation of pit heaps', *Proceedings of the University of Newcastle upon Tyne Philosophical Society*, Vol. 1, No. 9 (1967), pp. 129–36.

H. L. Ripley, 'Analysis of latest experiments in wet tipping at Egham', *Municipal Journal*, Vol. 68 (1960).

C. W. Rowell, 'Airfields into arable', *Farmers' Weekly*, 5 June (1959).

Sir E. J. Russell, 'Agricultural restoration of mining and quarrying sites', *Agriculture*, Vol. 54 (1947).

——, 'Rehabilitation of devastated areas', *Journal British Association for the Advancement of Science*, Vol. 7 (1951).

Sand and Gravel Association, *Pit and Quarry Textbooks* (MacDonald & Co., 1967).

M. Smith, 'The survey of derelict land', *Journal, Town Planning Institute*, Vol. 45 (1959).

L. D. Stamp, 'The reclamation of abandoned industrial areas', *Journal, Royal Society of Arts*, Vol. 100 (1951).

M. W. Thring (ed.), *Air Pollution* (Butterworth, 1957).

United States Soil Salinity Staff, *Diagnosis and Improvement of Saline and Alkali Soils* (Handbook no. 60, USDA, Washington, 1954).

United States Soil Survey Staff, *Soil Survey Manual*. Handbook no. 18 (USDA, Washington, 1951), 205–33.

University of Newcastle, *Landscape Reclamation*, Vols 1 and 2 (IPC Business Press Ltd, 1971/2).

W. F. J. van Beers, *Acid Sulphate Soils*. Bull. 3 Int. Inst. Land Recl. and Imp., Wageningen (1962).

C. J. Vyle, 'Landscape planning approach to land reclamation', *J. Inst. Parks Admin.*, Vol. 33, No. 1 (1968), pp. 22–4 and Vol. 33, No. 2 (1968), pp. 37–42.

R. O. Whyte and J. W. B. Sisam, 'The establishment of vegetation on industrial wasteland', *Common W. Agr. Bur. Jt. Publ. No. 14* (Aberystwyth, 1949).

W. O'N. Wilde, 'Land restoration of ironstone workings', *Chartered Surveyor*, Vol. 94 (1961).

P. Wix (ed.), *Town Waste Put to Use* (Cleaver-Hume, 1960).

R. F. Wood and J. V. Thirgood, 'Tree planting on colliery spoil heaps', *Forestry Res. Paper*, Vol. 17 (1955).

S. W. Wooldridge and S. H. Beaver, 'The working of sand and gravel in Britain', *Geographical Journal*, Vol. 115 (1950).

Working Party, *Taken for Granted*, Report on Sewage Disposal (HMSO, 1970).

———, *Pollution: Nuisance or Nemesis?* (HMSO, 1972).

J. C. Wylie, *Fertility from Town Wastes* (Faber and Faber, 1955).

Yorkshire Advisory Committee on Mining Research: 'Extinction and prevention of colliery spoil heap fires', *Transactions, Institution of Mining Engineers*, Vol. 101 (1942).

Chapter 8

Pollution of the Seas

by PAULINE K. MARSTRAND

Senior Research Fellow, Science Policy Research Unit, University of Sussex

In connection with the sea, pollution has been defined [1] as substances entering the sea and estuaries as a result of human activity causing harmful changes to the water, the ecology or the fisheries, or being detrimental to amenities such as bathing beaches, swimming areas or beauty of coast lines. The seas, viewed from most land-based vantage points, appear limitless, covering as they do, about two-thirds of the surface of the planet, about 140 million square miles. Small wonder that until very recently the capacity of this huge body of water to dispose of waste was assumed to be almost infinite. Why, within the last few years, has speculation been rife concerning the ability of the seas to go on removing the garbage of life, and even on the advisability of continuing to use the seas in this way?

Until well into the present century the effects of human activity on the sea were always localised and apparently fairly transient. Sometimes the sewage of estuarine towns slopped up and down in the estuaries without being carried away, as indeed it still does in some places. Excessive growths of *Enteromorpha sp.* were noted off Belfast and some other towns [2]. Occasional outbreaks of food poisoning were associated with shell-fish gathered near sewage outfalls, but most of the pollution was invisible, sporadic or not particularly unpleasant. In the 1930s oil began to be a nuisance on bathing beaches near to shipping lanes, such as the south coast of Britain, and blooms of algae and dinoflagellates were observed on the eastern US seaboard. Attention from these was however diverted by the war and the main concern with marine pollution has developed since 1945.

Britain, parts of Europe and the United States began to enjoy a boom in prosperity in the late 1940s and early 1950s. More people than before went away for holidays, at first to resorts in their home country, but increasingly to foreign tourist centres. In both cases the arrangements for disposal of sewage were rudimentary, pipes discharging as the tide was about to ebb, and the unplanned arrival of ever larger numbers of tourists led to the exposure of the inadequacy of these arrangements. Closed seas like the Baltic and the Mediterranean were particularly vulnerable and the unpleasant results forced many resorts to adopt at least primary treatment by screening and sedimentation before discharge. The resulting sludge was usually disposed of by dumping from barges or by releasing at night when the tourists were abed.

Where off-shore currents were strong, the flow of sewage was less obvious, but many of these places began to experience increasing pollution by oil, spoiling clothes and belongings. Public outcry about this in several countries led to regulations [3] about cleaning of oil tankers in territorial waters, but these were difficult to enforce and in any case did not prevent the occasional massive slick resulting from collision or accidental spillage. The increased demand for oil and its products brought more frequent sailings by larger and larger tankers and another unpopular result was the increasing numbers of oiled sea-birds found on beaches. It was obvious that discharge by tankers cleaning at sea was still producing considerable amounts of oil. In 1969 [4] the Intergovernmental Maritime Consultative Committee (IMCO) recommended the adoption of a procedure known as 'Load on Top', whereby the tanks are washed in succession, the oil being allowed to float up. Finally, all the oily washings are in one tank. The water, containing a small amount of oil is allowed out at a rate not exceeding sixty litres per mile of travel, while the oily waste with some entrained oil is retained in the tank. The next cargo is loaded on top of this. It is estimated that about two million gallons per year have been prevented from entering the sea by the adoption of this method by 80 per cent of the world's oil tankers, but the 20 per cent operating under flags of convenience do not comply, and the total amount of oil being shipped has increased so rapidly that the improvement has passed almost unremarked.

As urbanisation and industrialisation proceeded, and popular aspirations rose, the general awareness of surroundings grew too, so that first in Britain, then in Continental Europe and America, there was a demand for cleaner water and air. From the late nineteenth century on this led to diversion of domestic and industrial waste to rivers, and later to treatment works which themselves discharge to rivers, while air pollution regulations, often requiring 'scrubbing' of flue gases with sprays of water, diverted noxious materials from the air to the rivers and eventually to the sea.

Historically the main conservation activity connected with the sea has concerned stocks of fish. Attention has focused on gauge of net rather than on possible damage due to pollution, and the realisation that fisheries might be harmed has become widespread only in the last two decades. Pollution harms fisheries in three ways. The fish may become inedible and therefore unmarketable, for instance, if they are tainted by oil or heavily contaminated with a poison such as mercury. Secondly the food organisms which form the diet of the fish may be reduced, so that the fishery declines, or, as seems to be happening off Devon [5], the discharge of waste may redress a previous deficiency. Lastly, pollution may destroy the breeding grounds of fish, by accumulation of finely divided solids or by poisoning the eggs or fry. In many cases pollution does not destroy all the organisms, but the changes in population which result may not be desirable to man. In many developing countries where fish from inshore waters are an important source of protein, rapidly developing industry with insufficient planning of pollution control can be disastrous, as with the development of mercury [6] and cadmium poisoning in Japan.

The living biomass of the seas is probably about 90 per cent phytoplankton—microscopic plants living in the surface layers, using energy from the sun to combine carbon dioxide, water and mineral salts into plant material and food. In doing this they not only 'fix' some 4 per cent of the incident energy for future use by other organisms, the only input to stored energy which the world as yet has, they also release oxygen, probably about 70 per cent of the total [8].

There is no direct evidence that pollution is harming these organisms, but it is very probable that changes are occurring, and if these were to result in reduced photosynthesis, all the rest of the biosphere would almost literally have its base kicked from under it. Until we understand more about ecosystems of the sea, especially of the central parts of oceans, which seems rather unproductive, it is necessary to proceed with great caution in any activity which could alter the plankton community.

As awareness of the problems has become more sophisticated, so has the appreciation of the methods needed to prevent them, and from simple bans on washing ships in coastal waters there has been a progression to demands for more fundamental research on the existing ecosystems of the sea in order to understand what is happening and how hazardous it is. Among the most recently realised factors is the interrelationship between sea and air. Many of the pollutants come mainly from the air. Also, the rate of increase of concentration of carbon dioxide in the atmosphere has been slower than predicted, and this could be due to its increasing solution in the sea and subsequent combination with dissolved carbonates to form bicarbonates. Too little is yet known about the capacity of the sea to store carbon dioxide derived, among other sources, from the burning of fossil fuels [9].

Sources and Types of Pollution

Pollution enters the sea as run-off from land in rivers or pipes, from the air in rain or by solution of gases, by spillage during transport and by deliberate dumping of substances no longer required. Cowell [10] has developed a six-point pollution scale against which to judge the hazard due to a particular pollution.

Scale of Pollution

1. *Environmental addition*
   (a) Aesthetic damage: View spoilt or place rendered unsightly but no detectable biological damage. For example, polythene bottles.
   (b) Possible threat: Substance detectable and shown to be a potential biological danger experimentally. Level below that at which biological change can be detected.
2. *Environmental contaminant:* Some biological change detectable but not considered to be serious.
3. *Environmental hazard:* Organisms at risk. Effect causes concern among ecologists and is sufficient to warrant action to reduce the level in the environment.
4. *Environmental pollutant:* Organisms die and the level of pollution is high enough to merit public concern. The Species Diversity Index* of the area is reduced with consequent loss of ecosystem stability.

---

* Species diversity is the relationship between the number of species constituting the community and the total number of individuals of all species in that community. It can usefully be expressed for all stated unit area as the Species Diversity Index (SDI). Several methods of calculating Species Diversity Indices are available, but one commonly used is calculated as follows (Odum 1959)

$$SDI = \frac{\text{Total number of species in the community}}{\sqrt{\text{Total number of individuals of all species in the community}}}$$

(a) Chronic pollutant: Continued low levels result in pro-gressive damage at a rate higher than that of recovery processes. Situation will deteriorate unless measures to improve the situation are taken. For example, industrial effluents.

(b) Acute pollutant: Damage may be sudden and serious but recovery will usually take place naturally. Usually found as a consequence of isolated incidents. For example, localised oil pollution.

5. *Dangerous pollution:* Official action is automatic.

(a) Biological damage: Severe, but recovery is possible if action is swift and effective. Mortality will be high, with species threatened over a wide area.

(b) Radio-activity: Mortality may be low initially but species endangered as a result of genetic damage. For example, radiation level more than 0·5 rad/year.

6. *Catastrophic or disastrous pollution:* Widespread heavy mortality with little or no hope of rapid recovery. Species may be eliminated locally or even nationally. For example, the aftermath of nuclear, chemical, or biological warfare.

Using results obtained by other workers he has arranged some of the well-known pollution sources on this scale, shown in Table 1.

Oil. Oil is derived from natural seepage [11], from offshore drilling operations, from spillage during transport, from coastal oil-refineries, and from the air in rain. Long-term low concen-trations are more harmful than sudden massive incidents, al-though the latter attract more public disapproval and have there-fore tended to be the subject of more research. Crude oil is less toxic than fuel oil, and the oil in run-off water from the world's rivers and rain contained an estimated 200,000 tons of lead [12] in 1971. Split oil can be oxidised by bacteria [13] or by auto-oxidation catalysed by mineral salts in the sea, but the rate at which it is broken down depends on temperature and is very slow in cold seas.

A low index is normal only at the early stages of successional processes. In climax situations, a high SDI is usually obtained, indicating stability. Changes in the ecosystem producing lowered indices are indicative of lowered eco-system viability (MacArthur 1955, Margalef, R. 1968).

Table 1. *Relative positions of common pollutants on the scale proposed by Cowell*
(Arrows indicate weighting towards higher or lower categories)

Some of the authorities used in allocating scale placings		Pollution source	1	2	3	4	5	6
O'Sullivan 1969	1	Polythene bottles	←X					
Crapp 1970, Nelson-Smith 1968a and b Cowell 1969	2	98% Milford Haven oil pollution spillages	X					
Holdsworth 1970	3	Enzymes from detergents	←X					
World Press	4	Oil pollution from tankers operating LOT*	X					
	5	Dumping of American nerve gas	X↑					
	6	Detergent burden in London drinking water	X↑					
Jernelöv 1968	7	Methyl mercury in UK		←X				
E. C. Raney and B. W. Menzel, 1967	8	Hot water from power and industrial plant		←X				
Crapp 1969, Baker 1969	9	Oil pollution from refinery effluents		←X				
Nelson-Smith 1970, Holdsworth 1970	10	Oil from tankers not operating LOT		X				
Scott Russell 1969	11	Present level of disposal of radio-active waste		X				
Cowell 1969, Horizon—Whose Coast? BBC TV	12	Tourist damage to the Pembrokeshire coast		X				
	13	Agricultural fertilisers in British rivers		X				
Cowell 1970 + Private communication	14	Tourist damage to the Cornish coast		X↑	↑			
	15	Organic effluents in the Thames		X↑	↑			
Bellamy	16	Suspended sediments from Durham coast			X			
Riseborough 1968	17	Chlorinated hydrocarbons in the Atlantic			←↓X			
Cowell 1969, Cowell and Baker 1969, Bourne 1970	18	Oil from *Chryssi P. Goulandris* accident 1967			←↓X			
Woodwell *et al.* 1967	19	Chlorinated hydrocarbons in UK and Europe			X	↑		
Riseborough 1968, Wurster & Wingate 1968	20	Polychlorinated biphenyls in Atlantic			X	↑		
	21	Sewage disposal in Forth Estuary			X	↑		
Wurster 1969	22	Chlorinated hydrocarbons in USA			X	↑		
Bourne 1970, Smith 1968, Ranwell 1968	23	Oil from the *Torrey Canyon* accident			X	↑		
Crapp 1969, Cowell 1969	24	Detergents used in cleaning *Chryssi P. Goulandris* oil			X	↑		
Abelson 1969, Jensen *et al.* 1969	25	Methyl mercury in the Baltic				←X		
Abelson 1969	26	Methyl mercury in Lake Erie, USA				X→		
Crapp 1969, Baker 1969, Nelson-Smith 1970	27	Detergents used after *Torrey Canyon*					←↓X	
Jensen & Jernelöv 1969	28	Methyl mercury, Minimata and Agano Rivers Japan					X	
Frisch 1970	29	Hiroshima and Nagasaki atomic explosions						←X
Wolfers 1970	30	World population explosion						X→

* 'Load on top' system for disposing of oily wastes at sea.

Chlorinated hydrocarbons. DDT and its derivatives, and aldrin and dieldrin have all been detected in the sea. The distribution of concentration suggests that they are first carried by wind and then enter the sea from the air [14]. The rest must enter as run off from land [15]. Woodwell [16] estimates that there was perhaps sixteen billion pounds of DDT in the sea by 1967. There is evidence of damage to marine birds and to fish, and food-webs, as yet little understood, are probably being disturbed in ways we have no means of detecting. Polychlor-biphenyls, derived from fire-resistant materials, plastics and paint manufacture, have also become dispersed globally [17], and seem to have effects similar to those of DDT. These compounds are all more soluble in fats than in water. In living organisms they therefore accumulate in the storage tissues, and as they are not very rapidly metabolised, the concentration in these tissues may be orders of magnitude higher than that in the surrounding seawater. Birds and fish which have died in unexpectedly large numbers have been found to have high concentrations of these substances (DDT and/or PCBs) in their fat, sometimes accompanied by lesions of the liver [18]. It is thought that during times of starvation or stress during storms the fat stores are metabolised, thus releasing sub-lethal amounts into the bloodstream of already weakened animals. The behavioural changes induced may then be sufficient to cause death. Shellfish seem to concentrate halogenated hydrocarbon to about 70,000 times that in the ambient seawater, according to Peterle [19]. It has been suggested that these substances could be successively concentrated up to trophic levels.

The concept of trophic levels as first articulated by Lindemann is described succinctly by Philipson [20]. If micro- and macroscopic plants are regarded as producers, then they are eaten by first-order consumers, these by second-order consumers and so on. At each stage some energy is lost due to inefficiency of transfer and some by respiration. The protoplasm of one stage is used by individuals in the next partly as a substrate from which to release energy by respiration and partly as material for building new organisms. Thus the organisms at each level have to consume more material than they actually utilise. If the material contains DDT or other substance metabolised only slowly or not at all, then the concentration of this substance will increase with each successive level until the 'top predator' consumes a very much larger concentration than the first-order consumers. These results seem to be verified by the high levels found in birds of prey [21] and in predatory fish such as the cod. Since people do consume marine 'top predators' there is a risk to human health. The reproductive failure of many birds of prey in Britain, such as the Peregrine, correlates well with the use of pesticides, and since the control of the use of these substances there has been a marked improvement in breeding success of these birds [22]. Assuming that the effects on human health are small, does the loss of a few species of birds really matter? Apart from their undoubted aesthetic appeal, birds of prey probably perform a very useful role in the control of populations of other animals, and are thus necessary to the maintenance of balance between populations. Penguins are fish eaters, and too little is known of antarctic ecosystems to predict the effects likely to ensue from drastic reduction in their numbers.

Bio-degradeable organic waste.   This includes all organic substances which can be broken down by bacteria. In this process oxygen is used, so although most of them are not toxic in themselves the effect on water may be very deleterious. Such waste is contributed by domestic sewage, agricultural waste products, the food industry and the pulp and paper industry. A small amount is contributed by sewage from ships, but most of it enters the sea from rivers or discharge pipes. Nowadays it usually includes detergents which are not completely degradeable and which also contain fillers such as phosphates, or nitrilotriacetates, whiteners such as borates, and various inert foaming agents, all of which can be harmful in some circumstances.

When domestic sewage is treated, most of the processes produce large amounts of non-toxic but very oxygen-demanding sludge—much of this is dumped in inshore waters, with identifiable effects on the organisms present.

Inorganic industrial waste.   Many industries produce solutions containing metal ions or acid ions which are not removed by conventional wastewater treatment, and until recently most of this was ultimately conveyed to the sea. Among the most dangerous are cyanide, sulphuric acid, copper, mercury, lead, arsenic, nickel, cadmium and antimony. It had been assumed that inorganic compounds of these ions would be adsorbed on to bottom deposits and thus be removed from circulation, but in 1969 Jensen and Jernelov [23] reported that bacterial ooze in the coastal waters of Sweden was capable of converting inorganic compounds of mercury into much more lethal alkyl compounds which are circulated through living organisms and are extremely toxic to vertebrate nervous systems.

Heat. Many industrial processes discharge heated cooling water, notably power stations. Because of the enormous volumes of water required, such processes are often sited on a sea coast or estuary and the heated water raises the temperature of the estuary. In temperate climates, where the range of temperature tolerated by indigenous organisms through the year is as much as 15°C, temperature rise alone may not be very serious, although certain species will be preferentially selected. However, the increased temperature increases the rate of bio-oxidation by bacteria and may result in lowered oxygen concentration with consequent death of fish and other animals. In Britain this effect has been reduced by the use of cooling towers which oxygenate the water before it is released and sometimes by the provision of weirs in estuaries or in discharge channels [24].

Warm water in an estuary or coastal area may encourage a change towards semi-tropical species brought in on hulls of ships. In the early stages of the development of such a community one or more organisms may become pests in the absence of their normal biological checks [25]. This could have unwanted long-term effects on the native flora and fauna.

Attempts have been made to utilise the warm water for the breeding of fish or shellfish. In Britain plaice have been reared [26] and prawns are under investigation [27] and in the US other crustaceans are reared. So far the results seem to show that such 'farms' can be successful, provided the stock can be grown on to marketable size under protected conditions. Young plaice released beyond the experimental area were subject to heavy mortality and did not make a significant contribution to subsequent year classes. They showed more variation in colour and marking than normal 'wild' populations, but it is not yet known whether this or aberrant behaviour led to increased capture by predators or whether some other factor is involved. The possibility of using heated cooling water before discharge to the sea deserves further research.

Deliberate dumping. Apart from sewage sludge, many other solid or semi-solid substances are dumped at sea: garbage, mineral washings, unwanted chemical residues, process sludges from industry, surplus chemical weapons, radio-active wastes. The philosophy, if the substance is unsightly but harmless, has been to take it out to the edge of the continental shelf and dump it, and if it might be harmful, to dump it in containers. In the first category the advent of non-degradeable plastics refuse has completely altered the situation, so that now plastic junk has been dredged up from even the deepest ocean, and almost every beach has its plastic driftline. These artefacts are unsightly but probably harmless, at least until the sheer volume becomes overwhelming. The potentially harmful substances are a very different case. Some have been dumped in metal drums [28], these eventually corrode, and although the theory is that they will do so slowly, they may in fact all fail at once, and have been known to release, for instance, mustard-gas and arsenical compounds as has happened in the Baltic. The more recent practice of enclosing such substances in concrete is also open to criticism, since this too erodes, albeit more slowly, and can also be cracked by impact, as for instance when rolled together by currents. Since very little is known about the movement of sub-surface currents it is not always possible to predict where the dumped material will be carried.

Radio-active waste is enclosed first in steel and then embedded in concrete. If nuclear fission expands greatly as a means of generating power, then the vast amount of waste for disposal will present serious problems, while the wastage of steel involved will be considerable unless other shielding materials are developed. The half-life of these low level wastes is on average twenty years, and with constant additions of more containers the seas will never again be free of them. When the containers break down, as ultimately they must, there will still be some radio-active material to be released [29].

Plastic fibres.   There have recently been complaints that floss-like fibres are being caught in drift nets, and when a plankton net is sampled numerous nylon and other fibres can be found among the microscopic organisms. Since these fibres will not decay there is cause for concern that the amounts are increasing and that the sources are not known. They could interfere with plankton feeders such as herring (and whales). Longer and coarser fibres, such as fishing lines, are a cause of nuisance to net fishermen and can have tragic results for larger marine species, as they do for land animals.

Hazards and Effects
Oil.   The effects of oil are among the best documented and have already been indicated. Although oil is eventually oxidised so that only hard tarry lumps remain it has been calculated by Zobell [30] that 1 litre of crude oil requires the oxygen dissolved in 400,000 litres of seawater, so that except in splash zones the process is very slow and deoxygenation and death to organisms can result.

Tainting of fish, fouling of beaches and destruction of plankton are among the most noticeable effects, while the effects on coastal and estuarine habitats of discharge of oil refinery waste over many years are probably the most harmful. In a sudden disaster incident, vegetation may be covered with oil, and the overground parts may die, but eventually the oil becomes oxidised and eroded away, and many plant species can grow again, either from spores or seeds or from underground organs [31]. If the pollution is constant, and particularly if it contains some of the lighter and more toxic fractions of oil, the whole plant population will eventually be poisoned. Relatively few species are hardy enough to exist under estuarine and coastal conditions, so when these go the soil is left bare. It is eroded by the sea and this causes changes in the coastline and in the pattern of sandbanks. At the least this is inconvenient for shipping, at worst it can lead to coastline erosion.

Oil drilling operations are another source of continuous pollution and will undoubtedly affect marine populations in areas under exploitation. Unless the existing composition of these populations is investigated before operations commence the opportunity to study their effects and perhaps to decide how hazardous they are will be missed.

Halogen hydrocarbons. The most serious potential hazard from these and related substances is the effect they may be having on food-webs. Until more is known of the interdependency of marine organisms it is impossible to tell whether or not the toxic chemicals already released have had an irreversible effect, or whether such effects are likely to be produced in the future. The need for research into marine populations in polluted and unpolluted seas is urgent for this reason. The other effects such as concentration in food animals and reduction in numbers of birds are also significant but more noticeable, and therefore already the subject of considerable investigation.

Because young stages of fish and shellfish are more susceptible than adults, inland waters and estuaries are in special need of protection from this kind of pollution.

Organic waste. The most harmful effect of this kind of essentially nutrient waste is eutrophication. The organic material is utilised by bacteria which in the process convert the protein content into nitrates. Phosphates and potassium salts are also present, sometimes in the former case augmented by detergent, and the resulting mixture forms a nutrient solution for plants.

Shallow waters are then characterised by excessive growths of seaweed, particularly the green *Enteromorpha spp.* but also *Ulva lactuca* the sea lettuce and fucoids such as *F. spiralis*. The two green algae are especially indicative of possible pollution by sewage. The rocks are very slippery at low tide, and during storms the weed will be broken off and may be deposited on bathing beaches, where its subsequent decay causes nuisance from smell and attracts flies which breed in the decaying mass. Similar effects occurred on the beaches which were cleared with detergent after the *Torrey Canyon* foundered, though in this case the excess growth was due to the almost total destruction of limpets which normally browse on the young plants [31].

If the water is deeper, phytoplankton blooms may occur. Some of the species favoured by these highly nutrient conditions produce toxins whch poison fish fry and food. Some dinoflagellates have caused 'red' and other coloured tides and seem capable of producing a toxin powerful enough to induce severe skin irritation in humans, and some get filtered out by oysters and other shellfish which are thus rendered toxic. When the bloom is over the decaying bodies of the phytoplankton may give rise to deoxygenation. Another major effect of sewage is suffocation and occlusion of light. The usually cloudy sewage slick screens light from plants, thus reducing the production of oxygen, and eventually settles on the bottom, blocking up crevices and making a habitat suitable only for animals such as *oligochaets* and burrowing molluscs which can survive in low oxygen conditions [32].

There has been considerable concern over the survival of human enteric pathogens in seawater. Evidence cited at a symposium in 1970 [33] suggests that few or none survive long when exposed to sunlight and air, but current investigations as part of the Sabrina project on the Severn estuary [34] indicate that much longer survival times are possible in some circumstances. In this case there is a real threat of spreading disease.

Inert solids. Finely divided non-degradable material occurs in domestic waste, and also in slurries pumped from mines and mineral processing plant. As a cloud in the water it occludes light and reduces photosynthesis. When it settles it clogs the bottom with an unstable unproductive ooze which eliminates many organisms. Comparison between affected and unaffected parts of the north-east coast of Britain shows the radical impoverishment of offshore water in these circumstances [35]. This reduction in diversity, if widespread, could initiate changes which would ultimately threaten fish stocks.

Inorganic waste.   Industrial waste containing metal ions is toxic to many forms of life. The conversion of inorganic mercury into alkyl compounds has indicated that substances previously thought to be removed from the biosphere by adsorption can be released as more harmful substances. Cadmium also has contaminated inshore fisheries in Japan [36], and it is probably only a matter of time before other substances, particularly metals, are identified as the cause of disease. This threat is particularly dangerous to developing countries, where the new industry may be introduced without effluent control plant, and where the infrastructure neces-sary for adequate control has in any case not been developed. Often the local population is dependent on the sea for the protein content of the diet, and with low levels of medical service the effects may not reach the attention of anyone who understands them for years, by which time hundreds of people will have been permanently affected and a valuable food resource rendered unuseable. The biological background has usually not been investigated beforehand and is difficult to interpret afterwards. In these situations it is imperative to ensure adequate control from the beginning, and if possible to assess possible side-effects before they occur.

Some of the effects of these compounds have been noted in the past from anti-fouling paints on boats [37], indeed many of these do contain mercury. With the increase in amounts of shipping entering harbours it will be necessary to regulate the amounts in the water, and possibly, as in Japan, to forbid the use of some anti-foulants, for ships using certain harbours.

Radio-active waste.   As yet the effects of radio-active waste are speculative rather than proven. Not much has escaped and natural radio-activity has produced communities presumably adapted to its continuous presence over thousands of years. The main poten-tial hazard is the increase in rate of mutation. Most genetic changes produce organisms which are not as well adapted to the given environment as the 'normal' individuals, and are therefore elimi-nated from the population by selection. These mutations occur at a more or less regular rate, so if the environment changes so that one or more of them are favoured, the population will change towards the new type. This is the basis of speciation. Radio-activity increases mutation rates, which could mean that disadvantaged organisms could be produced faster than they could be 'bred out' and it also induces faults in the genetic material which are often lethal or severely damaging. In micro-organisms, with a short generation-time, new varieties can be produced very quickly—for instance slime-forming bacteria in the cooling-tanks at Windscale seem to be producing new strains, resistant to slimicides, at a faster rate than usual [38]. Organisms could be produced which were resistant in this way, or which were not susceptible to normal checks. As the normal checks are often not known it is difficult to predict what the effects of long-term exposure to low levels of radio-activity might be. Many chlorinated hydrocarbons are also capable of inducing mutation and the possibility of combined effects has to be considered [39].

Fisheries. The most productive fisheries of the world seem to occur where the elevated ocean floor causes upwelling of currents bringing nutrients which support a variety of plant and animal plankton. Often the fish are not bred in the areas where they are caught, and many seem to pass their early years in coastal regions. For instance, the North Sea flatfish breed on the coasts of Belgium, the Netherlands, Germany and Denmark. Increasingly such areas are contaminated by polluted rivers or by discharge made directly to the sea. Since the eggs and larval stages of fish and their food organisms are especially sensitive to toxic substances, the fish stocks are threatened [40]. It is likely that the fishing nations are only just beginning to feel these effects because the development of most of the critical coastline is not yet complete, even in Europe, and because improved technique has kept up the total catch. There is however a marked decrease in size of some fish, such as cod, which may be due either to pollution or to overfishing or to a combination of the two. FAO has recommended special measures for the protection of inshore waters in order to protect breeding stocks [41] and the siting of new industry should be undertaken with such safeguards designed into the plant from the start. Once a breeding stock has been damaged it could very easily drop to below replacement level before the damage was appreciated, especially in areas where fisheries expertise is not yet well developed.

Mature biological communities are characterised by species diversity. That is there are a large number of different species each occupying a fairly well-defined ecological niche [42]. Small changes in the environment will produce little effect, because adjustments in relative numbers will occur in response and the community as a whole will appear remarkably stable. In the presence of an ecological disturbance, such as pollution, some of the species will be more vulnerable than others and will tend to disappear. Others, not themselves vulnerable, but dependent on the damaged species for food or shelter will also be eliminated. The whole community will be simplified to one with large numbers of a few species rather than small numbers of many. The ratios of various species to each other will also be changed. In freshwater, where ecological systems have been more extensively studied, systems have been developed to use these changes to assess the extent of pollution. Regier and Cowell [43] suggest that similar schemes be developed for marine habitats when records of undisturbed waters have been made in many different kinds of sea.

The simplest form of diversity index is usually expressed thus:

$$DI = \frac{\text{Number of species present}}{\sqrt{\text{Total number of individuals}}}$$
$$\text{of all species}$$

and this gives a fairly quick 'rule of thumb' guide. However, it does not reflect changes in relative numbers of different species, which may well be indicative of particular kinds of pollution, and might be more useful if combined with the concepts included in the 'score' system of Chandler [44] for freshwater. In this system tolerant organisms found over a wide range of conditions are given a constant score—for freshwater around forty. Very sensitive species are given higher scores, increasing with increased sensitivity and with increased numbers. Species with adaptations to very polluted conditions, such as special respiratory mechanisms or behaviour, are given low scores, which decrease as their numbers increase. Table 2 illustrates the 'score' system.

As more is learned of the biota of polluted and unpolluted water, examination of the biological community will indicate whether and how much pollution has occurred, and, at least for some substances, what kind of pollution.

Measures to Control, Contain or Remove Pollution

Until recently nothing was usually done until a fairly extensive and obviously damaging incident occurred, and so measures to remove the offending substance have usually been the immediate response. As these have become increasingly forlorn or even disastrous there has been a welcome move towards preventive measures, including legislation. Whatever the measures taken,

Table 2. *Biotic index by the 'score' system*

Group present in sample	Increasing abundance				
	Pre-sent	Few	Com-mon	Abun-dant	Very abun-dant
	Points scored				
Each species of *Planaria alpina* Taenopterygidae Perlidae, Perlodidae Isoperlidae, Chloropedidae	90	94	98	99	100
Each species of Leuctridae, Capniidae Nemouridae (excl. Amphinemura)	84	89	94	97	98
Each species of Ephemeroptera (excl. Baetis)	79	84	90	94	97
Each species of Cased caddis, Megaloptera	75	80	86	91	94
Each species of Ancylus	70	75	82	87	91
— Rhyacophila (Trichoptera)	65	70	77	83	88
Genera of Dicranota, Limnophora	60	65	72	78	84
Genera of Simulium	56	61	67	73	75
Genera of Coleoptera, Nematoda	51	55	61	66	72
— Amphinemura (Plecoptera)	47	50	54	58	63
— Baetis (Ephemeroptera)	44	46	48	50	52
— Gammarus	40	40	40	40	40
Each species of Uncased caddis (excl. Rhyacophila)	38	36	35	33	31
Each species of Tricladida (excl. *P. alpina*)	35	33	31	29	25
Genera of Hydracarina	32	30	28	25	21
Each species of Mollusca (excl. Ancylus)	30	28	25	22	18
Chironomids (excl. *C. riparius*)	28	25	21	18	15
Each species of Glossisiphonia	26	23	20	16	13
Each species of Asellus	25	22	18	14	10
Each species of Leech, excl. Glossisiphonia, Haemopsis	24	20	16	12	8
— Haemopsis	23	19	15	10	7
— *Tubifex sp.*	22	18	13	12	9
— *Chironomus riparius*	21	17	12	7	4
— Nais	20	16	10	6	2
Each species of Air-breathing species	19	15	9	5	1
No animal life			0		

considerable expense is usually involved, either in the clearing up operation, or in the provision of treatment plant and control functions, and if developing countries are to be able to avoid the worst experiences suffered by those already developed, adequate financial backing will have to be provided on an international scale.

Oil. As one of the earliest and most visible objects of public complaint, the measures used to combat oil pollution have been until recently, largely remedial. There are five categories of this activity:

Containment—with wooden or inflated plastic booms [45].

Collecting —with straw [46], or more recently with skim pumps, 'slick lickers' or mineral wool fibre.

Sinking —with chalk or sand covered with a hydrophobic substance [47].

Dispersal by detergents and churning [48].

Degradation by bacteria [49].

The first is useful for keeping an oil slick out of a river mouth or inlet, and has been used to contain leakage from an oil well, but is not always reliable in rough seas and is best regarded as an interim measure, or as a preliminary to skim-pumping or slick-licking. The latter process consists in rotating a belt coated with absorbent material into the slick and then back into a cleaning tank. It has been used very successfully on inland lakes in Sweden and more recently for a marine spill in Canada.

Where time and labour permit, collecting is the best method, and if no pumps or lickers are available straw and peat are still effective absorbents, which can afterwards be burnt. The beaches of Brittany were cleared in this way after the *Torrey Canyon* incident in 1967 [50].

If the oil can be absorbed on to sand or chalk the oil mass will then sink. The French authorities had considerable success with 'Craie de Champagne' after the *Torrey Canyon* incident [51]. The chalk is covered with a hydrophobic substance so that it effectively segregates the oil from the water. There was some concern at the time that the sunken oil would damage fisheries, or cause nuisance by being brought up in nets, but this does not seem to have happened. It would probably be advisable to use aircraft to trail the slick and report when it was approaching an area where sunken oil would cause little or no damage before adding the chalk from ships. However, the success of the method depends on action while the slick is still fairly small and concentrated and before partial evaporation and wave action have turned it into the notorious water-in-oil emulsion known as 'chocolate mousse' which is the most difficult substance to remove from boat hulls and beaches.

Dispersal is a fairly logical remedy, especially as it presumably assists both bio- and auto-oxidation. Mechanical dispersal is difficult with large slicks, although the UK government laboratory at Warren Springs has developed a 'gate' which can be lowered off a tug and towed through a slick to churn it up [52]. It was fairly logical also to suppose that detergents would assist the dispersal process and many petrochemical companies developed special detergents. The use of these during the cleaning-up operations following *Torrey Canyon* and other spills between 1966 and 1969 showed however that they were extremely toxic to many marine organisms. The companies have been carrying out research into the reasons for this toxicity and several less toxic products are now on the market. Unfortunately, in the UK and possibly other maritime countries, the local authorities likely to be affected had already purchased quantities of the earlier products, and these are still being used.

Left to itself, the lighter and most toxic fraction of oil evaporates. The heavier constituents are separated into lumps by wave action, and may form blobs of 'chocolate mousse'. In the open ocean these blobs are slowly oxidised by biological and chemical processes and may finally be degraded into hard tarry lumps which sink. On-shore this process is inconveniently slow, and research is being undertaken at the University of Wales, Cardiff, to identify the effective bacteria and to develop methods of encouraging them [53]. The sequence of events seems to be that if crude-oil is deposited, the drying action of the air combined with bacterial and auto-oxidation render it into relatively dry, though still plastic lumps, which become progressively smaller and fewer over a period of four or more years. If the initial deposit was 'chocolate mousse' the much softer emulsion penetrates between stones and sand, and tends to form conglomerates which break, releasing the oil when trodden or sat upon. The break-down process proceeds in the same way, and might be expected to be quicker, because the smears of oil are thinner. In both cases the results are very unpleasant on bathing beaches. The Field Studies Council Oil Pollution Research Unit at Orielton (UK) has suggested a scale of urgency for beach clearing based on the use to which the beach is put [54]. If it is a bathing beach for tourists and if off-shore currents will not cause pollution of fisheries, use of detergents can be allowed.

If it is a bathing beach, but fisheries might be harmed, only mechanical removal should be attempted.

If it is a secluded part of the coast not often visited and the oil will not be lifted off by the tide, leave it.

Pesticides. In the long run the effects of pesticides and other organic chemicals may be the most serious hazard. Apart from avoidance of spillage and control over effluents from manufacture and formulation plant, both relatively minor sources, the only way to reduce the amounts entering the seas may be to reduce their use in agriculture and medicine. In these circumstances very detailed investigation will need to be undertaken to determine how important such use is in any given region, economically, and socially, and how much damage is likely to result and where. Only in this way will it be possible to approach a situation in which correct decisions on whether or not to use these substances can be made [55].

Some countries have already banned or restricted the use of one or more persistent pesticides, e.g. Denmark, Hungary, Sweden, UK, US, and no doubt others will do so, but since it seems that windborne air is the principal source of contamination, international agreement will be necessary. This will not be achieved unless governments can have sufficient evidence to show that it is really necessary. As a result of the discussions before and during the Stockholm Conference it is to be hoped that all those who can will devote considerable resources to collecting the relevant information and evaluating the situation on a continuing basis.

An alternative method of control would be by barring the manufacture of certain compounds, but apart from possibly disastrous effects on developing countries in the short run, this would be extremely difficult to police, and could produce a 'black market' in some substances. However, some success has been achieved, the sole manufacturers of PCBs in UK have barred them for all but 'captive' uses, that is those where the substances are accounted for all the way through the process and not allowed to escape. This would be possible for compounds used in manufacturing processes, though not for those such as pesticides where use implies spread. Guidelines for use, such as those proposed in the concept of 'economic threshold' of use [56] on agricultural crops, could do much to reduce the levels at large in the environment.

Nutrients and heat. Both these forms of pollution are potentially useful. In developing countries it should be increasingly possible to site and design industrial plant so that surplus nutrients could be reclaimed and used in agriculture or aquaculture, in some cases associated with temperature increases. Technically it is already possible to use surplus heat for various purposes and others could be developed. Given adequate research followed by application of research results, the problems associated with these two could be held level or even reduced.

Harbours. Prevention of pollution of harbours requires agreement between all countries whose ships use them. Nations can only exercise their right to refuse entry to ships with, for instance, oily discharges, or toxic anti-foulant paint if avoidance is technically and economically possible. If ships are to use 'load on top' and retain some contaminated water, harbour facilities will need to include holding tanks for this water and subsequent removal or separation.

Financial backing. To be effective, measures taken to control any source or type of pollution will have to be agreed by all the nations which use the part of the oceans affected. The twelve-nation agreement on dumping of waste in the North Sea and North East Atlantic [57] is an example of the kind of regional control which can be achieved, and the further measures proposed by IMCO on navigation to reduce danger of collision are the basis for a wider form of agreement. Where one or more of the nations concerned are economically unable to meet the requirements of a proposal, an international body, such as the environmental 'organ' agreed to at the UN Conference on the Human Environment, Stockholm, 1972, should be able to direct funds to assist such countries. After all, every country ultimately loses if measures to protect the seas are not effective.

International recommendations. The bases for agreement have in many cases already been put forward by international bodies (see Chapter 9). The 1954 convention on prevention of pollution of the sea by oil was amended in 1969 to include adoption of 'load on top'. The US have already proposed an agreement to ban drilling beyond the 200 metre isobath. The International Atomic Energy Agency has issued guides for the transport and disposal of radioactive waste [58]. The Group of Experts on the Scientific Aspects of Marine Pollution, (GESAMP) has made broad recommendations on all forms of marine pollution [59] while UN Food and Agricultural Organisation and World Health Organisation have drawn up a course in coastal pollution control.

Some port and dock authorities have drawn up contingency plans for dealing with pollution and have developed policies for managing their activities in order to minimise its effects [60].

Monitoring. None of the control measures can be effective without a deeper understanding of the ecology of the sea and without constant vigilance. Tarzwell [61] has suggested standardisation of toxicity tests so that results can be comparable and comprehensible to all nations and others are devising methods of using coastal and estuarine communities as monitors [62, 63].

The pre-Stockholm International Working Group on Marine Pollution stated as one of the objectives of marine pollution control the identification of critical substances; those whose presence, even in small amounts, was likely to cause either effects which would not become obvious until considerable damage had been caused, or which would cause serious and lasting damage. Research to achieve this goal would do much to assist in applying the other control measures, and in the determination of priorities. The group also suggested that it should be made obligatory to obtain a licence before allowing any 'critical' substance to enter the sea. An international register of types, amounts and position of release of all substances would be a useful adjunct for any international control programme. The necessary evaluation of the pesticide problem would be easier had such a register been kept in the past.

The open sea. Parts of the seas remote from any coastline are only recently coming into the research purview. This research will need to be stepped up in order to collect background data before the seas become further altered by human activity.

The absorptive capacity of any country is becoming an important natural resource. The considerable absorptive capacity of the oceans is an international resource of immeasurable value which must be protected by the combined efforts of all nations, singly, in groups, and through UN. Probably it is very far from saturation, but as there is no means of being certain, it will be wise to proceed cautiously, remembering that it must ultimately have a limit for things which cannot be degraded, and that for those which can, there is a delicate balance to be preserved.

References

1. UN Conference on the Human Environment, *Intergovernment Working Group on Marine Pollution*, Report A/Conf. 11/5 (1972).
2. E. J. Perkins and O. J. Abbot, 'Nutrient enrichment and sandflat fauna', *Marine Pollution Bull.*, Vol. 3, No. 5 (May 1972).

3. 1954 Convention for Prevention of Pollution of the Sea by Oil.
4. As above, amended 1969, 9 Int. Legal Materials 1 (January 1970).
5. F. S. Russell *et al.*, 'Changes in biological conditions in the English Channel during the last half century', *Nature*, Vol. 234, No. 5330 (24 Dec. 1971), pp. 468–70.
6. Kumamoto University, Japan, 'Minamata disease', summary of papers by Minamata study group from 1958 to 1968.
7. K. Tsuchiya, in *Archives of Environmental Health*, Vol. 14 (1967), p. 875.
8. O. Schachter and D. Serwer, 'Marine pollution problems and remedies', *AJIL*, Vol. 65, No. 1 (Jan. 1971).
9. A. V. Kneese, 'Background for the economic analysis of environmental pollution', *Swed. J. of Economics* (1971).
10. E. B. Cowell, 'Oil pollution in perspective', in Cowell (ed.), *Ecological Effects of Oil Pollution on Littoral Communities* (Institute of Petroleum, 1971).
11. J. Hahn, in 'Natural oil seepage', XV *Oceanus* 12 (Woods Hole Oceanographic Inst., October 1969).
12. Picard 1971, quoted in n. 8 above.
13. S. Beastall, unpublished work in progress at Univ. Coll. of S. Wales, Cardiff, in Dept of Microbiology.
14. R. W. Riseborough, 'Chlorinated hydrocarbons in marine ecosystems', in Miller and Berg, *Chemical Fallout: Current Research on Persistent Pesticides* (Springfield, Illinois, Chas. G. Thomas, 1969).
15. C. F. Wurster and D. B. Wingate, 'DDT residues and declining reproduction in the Bermuda petrel', *Science*, Vol. 159 (1968), pp. 979–81.
16. G. N. Woodwell, 'Toxic substances and ecological cycles', *Scientific American*, Vol. 216 (March 1967), p. 24.
17. R. W. Riseborough *et al.*, 'Polychlorinated biphenyls in the global ecosystem', *Nature*, Vol. 220 (1968), p. 1098.
18. M. W. Holdgate (ed.), 'The Seabird wreck of 1969 in the Irish Sea', Report by Natural Environment Research Council (London, 1971).
19. T. J. Peterle, 'Pyramiding damage', *Environment*, Vol. 11 (July–August 1969), p. 34.
20. E. J. Philipson, *Ecological Energetics*, Studies in Biology No. 1 (Edward Arnold, 1966).
21. N. W. Moore, 'Pollution by pesticides', Proc. of Conference on Pollution, Liberal Party, London in November 1970.
22. I. Prestt, D. J. Jefferies and N. W. Moore, 'Polychlorinated biphenyls in wild birds and their avian toxicity', *Environmental Pollution*, Vol. 1 (1970), pp. 3–26.
23. S. Jensen and A. Jernelov, 'Biological methylation of mercury in aquatic organisms', *Nature*, Vol. 223 (1969), p. 753.
24. R. S. A. Beauchamp, in Paper to Thermal Workshop convened in US by the International Biological Program (3–7 November 1968).
25. Cowell, in Cowell, op. cit.
26. Coughlan and Spencer, 'Power station and aquatic life', in *Effects of Industry on the Environment* (Field Studies Council, 1970).
27. New Rank Research Organization, High Wycombe, Bucks. Work in progress.
28. Anon., 'Plight of the Baltic', *Marine Pollution Bulletin*, Vol. 1 (4) (1970).
29. R. Scott-Russell, 'Contamination of the biosphere with radio-activity', Vol. 2, No. 1 (1969), pp. 2–9.

30. C. E. Zobell, 'The occurrence, effects and fate of oil polluting the sea', Proc. Int. Conf. on Water Pollution Research (Pergamon, 1964).
31. J. M. Baker, 'The effects of a single oil spillage', in Cowell, op. cit.
32. F. Beyer, 'Zooplankton, zoobenthos and bottom sediments as related to pollution and water exchange in Oslofjord', *Helgolande wiss. Meeresunters*, Vol. 17 (1964), pp. 496–509.
33. A. L. H. Gameson, D. Munro and E. B. Pike, 'Effects of certain parameters on bacterial pollution at a coastal site', *Water Pollution Control in Coastal Areas*, Proc. of Symposium held by Inst. Water Pollut. Control, Bournemouth, May 1970; *Water Pollution Control*, Vol. 69, No. 4 (1970), pp. 34–51.
34. G. C. Ware, A. E. Anson and Y. F. Arianayagan, 'Bacterial pollution of the Bristol Channel', *Marine Pollution Bull.*, Vol. 3, No. 6 (June 1972), pp. 88–90.
35. D. J. Bellamy and A. Whittick, 'Problem of the assessment of the effects of pollution on inshore marine ecosystems', in *The Biological Effects of Pollution on Littoral Communities* (Field Studies Council, Supplement 2, 1968).
36. K. Tsuchiya in *Archives of Environmental Health*, Vol. 14 (1967), p. 875.
37. A. J. O'Sullivan, 'Marine Pollution' paper to 13th Annual Conference of Inst. Water Pollution Control (Brighton, 1971).
38. Ripon, CEGB., unpublished work.
39. B. Bridges, 'Environmental genetic hazards', *Ecologist*, Vol. 1, No. 12 (June 1971).
40. P. Korringa, 'Biological consequences of marine pollution', in *Helgolandewiss, Meeresunters*, Vol. 17 (1–4) (Hamburg, April 1968).
41. FAO. Report of technical conference on marine pollution and its effects on living resources and fishing (Rome, 9–18 Dec. 1970).
42. Cowell, in Cowell, op. cit.
43. G. B. Crapp, 'Ecological effects of stranded oil', in Cowell, op. cit.
44. J. R. Chandler, 'A biological approach to water quality management', *Water Pollution Control*, Vol. 69, No. 4 (1970).
45. D. E. Newman and N. I. Macbeth, 'The use of booms as barriers to oil pollution in tidal estuaries and sheltered waters', in P. Hepple (ed.), *Proc. Seminar on Pollution of Water by Oil* (Inst. Wat. Pollut. Control and Inst. Petroleum, Aviemore 1970).
46. L. Cabioch, 'The fight against pollution by oil on the coasts of Brittany', ibid.
47. O. G. Mironov, 'Hydrocarbon pollution of the sea and its influence on marine organisms', in *Helgolande wiss. Meeresunters*, Vol. 17 (1–4) (Hamburg, April 1968).
48. J. Wardley-Smith, 'Methods of dealing with oil pollution on and close to the shore', in Hepple, op. cit.
49. G. D. Floodgate, 'Microbial degradation of oil', *Marine Pollution Bull.*, Vol. 3, No. 3 (March 1972).
50. J. E. Smith (ed.), *'Torrey Canyon' Pollution and marine life*, Report by Plymouth Laboratory (Cambridge UP, 1968).
51. Mironov, op. cit.
52. See 48.
53. See 49 and 13.
54. J. M. Baker and G. B. Crapp, 'Predictions and recommendations', in Cowell, op. cit.

55. G. R. Conway, 'A consequence of insecticides' in M. Taghi Farvar and J. Milton (eds), *The Unforeseen ecological boomerang* (Nat. Hist. Special Supplement), 46.
56. US Senate 'Pesticides and public policy', Report No. 1379 to 89th Congress.
57. *Financial Times*, 22 January 1972.
58. IAEA Guides for Transport and Disposal of Radio-active Wastes. Safety Series, numbers 5, 6, 11 and 27.
59. GESAMP, Report of second session of Joint Group of Experts on Scientific aspects of marine pollution, GESAMP II/11 (20 June 1970).
60. Capt. G. Dudley, 'Oil pollution in a major oil port', in Cowell, op. cit.
61. C. M. Tarzwell, 'Toxicity of oil and dispersant mixtures to aquatic life', in Hepple, op. cit.
62. G. B. Crapp, 'Monitoring the rock shore', in Cowell, op. cit.
63. H. A. Regier and E. B. Cowell, 'Applications of ecosystem theory, succession, diversity, stability, stress and conservation', *Biological Conservation*, Vol. 4 (2) (January 1972).

Chapter 9

The Law Relating to the Pollution of the Seas

by THOMAS A. MENSAH

Head of Legal Department, International Maritime Consultative Organisation

Introduction

Legislation on the prevention and control of pollution of the sea must, if it is to cover the subject fully, take account of two very important characteristics of marine pollution. The first of these is that marine pollution does not arise solely from activities in the marine environment. Pollution of the seas and oceans can and does result from activities which take place in other sectors of the human environment: from land and from the atmosphere. Any programme designed to regulate and control marine pollution must, therefore, have some regard to activities in these other sectors which may have some polluting effects on the sea. The second important characteristic of marine pollution is that, owing to the way in which the seas and oceans bind together the countries and continents in an unbroken and a multi-directional link, it is usually difficult, and in some cases impossible, to isolate completely the effects of pollution incidents within the territorial confines of single states. For these reasons the control and prevention of marine pollution has for the most part been regarded as an international rather than a national matter, and many of the measures and schemes in this field have been devised through international discussion and co-operation. As a result it is possible to discover a wide measure of international agreement, not only in the objectives of governments as expressed in international treaties and similar instruments, but also in the actual content of the legislative measures adopted by various states to deal with the problem.

As with other forms of pollution, marine pollution is of two main kinds. First there is pollution arising from the general operation of the environment (including the other sectors such as land and the atmosphere). This is 'natural' pollution. Then there is pollution arising from various human activities which directly or indirectly have a polluting effect on the environment. This is man-created pollution. The legal control of marine pollution (as indeed of pollution of the other sectors of the environment) has, hitherto, been viewed primarily in relation to man-created pollution. While this emphasis is doubtless understandable and, perhaps, even necessary, it is well to bear in mind the importance of 'natural pollution' and its relevance to the overall success of any programme to preserve the environment.

Marine pollution arising directly or indirectly from human activity can be considered under a large number of different category headings depending on context. For the present purpose it is sufficient to deal with it under three main category headings, namely:

1. Pollution arising from the disposal (whether deliberate or through accidental escape) of domestic and industrial wastes into the sea.
2. Pollution arising from the use of the marine environment by ships and the consequential discharge or accidental escape of shipborne pollutants.
3. Pollution of the sea arising from activities connected with the exploration of the area of the sea-bed and ocean floor and subsoil thereof and the exploitation of the resources of this area.

Each of these types of pollution requires more or less different measures for its prevention and control. Some of the measures are taken at state level, others are taken (and require to be taken) at the international (global, regional or sub-regional) level, or at least through inter-state co-operation and agreement. Yet others require a combination of national and international action at various levels. In all cases the measures include, in addition to various practical and technical measures devised and improved from time to time by science and technology, legal measures of restraint and sanction as well as administrative arrangements for applying such measures, evaluating their efficacy and, as necessary, improving them in the light of changing circumstances and felt needs. These legal regimes vary, in objective and content, from state to state and according to the type of pollution they are meant to regulate and control.

United Kingdom Legislation

Discharge of domestic and industrial wastes [1]. In England and Wales the discharge of domestic and industrial wastes (effluents) is governed by a number of Acts and subordinate legislation made thereunder. The principal Acts are:

(a) The Rivers (Prevention of Pollution) Acts, 1951 and 1961.
(b) The Clean Rivers (Estuaries and Tidal Waters) Act, 1960.
(c) The Water Resources Act of 1963.

In Scotland the equivalent Acts are the Rivers (Prevention of Pollution) (Scotland) Acts of 1951 and 1965.

The Water Resources Act establishes river authorities and endows them with power to exercise certain functions under the Rivers (Prevention of Pollution) Acts of 1951 and 1961 and the Clean Rivers (Estuaries and Tidal Waters) Act of 1960.

The Rivers (Prevention of Pollution) Acts control the discharge of effluents into rivers (streams). The Act of 1951 provides that it is an offence,

'for any person to cause or permit to enter into any stream any poisonous, noxious or polluting matter or to cause or permit to enter into any stream any matter so as to tend either directly or in combination with similar acts (whether his own or another) to impede the proper flow of the waters of the stream in such a manner leading or likely to tend to a substantial aggravation of pollution due to other causes or its consequences.' [2]

The Minister of Housing and Local Government (now the Secretary of State for the Environment) and various river authorities are granted divers powers, including, in particular, the power to prevent the use of any stream in a manner that would contravene the provisions of the Act. River authorities are also empowered to make bye-laws for the purpose of preventing pollution of the waters of the rivers within their jurisdiction. In addition river authorities are also empowered to authorise or refuse permission to any person within their jurisdiction 'to bring into use any new or altered outlet for the discharge of trade or sewage effluent to a stream or to make a new discharge . . .'. [3] Under the Act, the Minister is given the power to extend the provisions of the Act to estuaries and coastal waters.

The Clean Rivers (Estuaries and Tidal Waters) Act extends the jurisdiction and powers of river authorities to deal with the discharge of effluents to 'tidal waters or parts of the sea within the seaward limits (as specified in the Act)' [4]. The river authorities are also empowered under the Water Resources Act to arrange to take samples of effluents entering into a river or water under their control and, when satisfied that such effluent can cause pollution, require the persons responsible to take preventive or remedial action. Any river authority may also require information from persons concerned [5]. Penalties are attached to failure to comply with orders so given or to provide required information as well as to the giving of false information.

The Royal Commission on Environmental Pollution is not completely satisfied with the legislative and administrative arrangements in respect of pollution of the estuaries which, in their view, remain 'more vulnerable to pollution than any other part of the British environment' [6]. They have made a number of recommendations for legal and administrative action to deal with this problem [7].

In addition, section 3(1) of the Oil in Navigable Waters Act, 1955 prohibits the discharge of oil or oily mixtures 'from . . . any place on land or from any appurtenances used for transferring oil from or to any place on land'. Similarly, sections 6 and 7 of the Radio-active Substances Act, 1960, regulate disposal of radio-active wastes into the sea and estuaries [8].

This arrangement is voluntary and involves the Ministry of Agriculture, Fisheries and Food and the Ministry of Agriculture and Fisheries of Scotland. Under this arrangement the appropriate Ministries are consulted on proposals to dump toxic chemical wastes into the sea. Normally, disposal is suggested in waters deeper than 200 fathoms beyond the edge of the Continental Shelf although in certain cases where it is safe to do so, agreement may be given to dumping of other less toxic substances in shallower waters.

The United Kingdom is one of the signatory states to the Oslo Convention to regulate dumping into the North-East Atlantic and there is now every likelihood that the Government will introduce legislation to make the controls of that Convention compulsory in respect of United Kingdom operators [9].

Dumping of wastes into the sea. There is no separate legislation against dumping of waste materials into the sea. Harbour and river authorities have power, under legislation relating to river pollution and water pollution, to make bye-laws to control the dumping of polluting substances into harbours, rivers, estuaries and tidal waters and other areas of United Kingdom territorial waters. No other statutory control exists, although there is an arrangement under which the government is consulted if special problems arise or if there is a danger of interference with fisheries as a result of the dumping of some substances.

One of the recommendations of the Royal Commission on Environmental Pollution is that 'legislation to implement the Oslo Convention on the Control of Marine Pollution by Dumping from Ships and Aircraft in the North East Atlantic should be introduced as a matter of priority' [10]. The Government has already indicated its intention to introduce legislation on the subject soon [11].

Pollution by shipborne pollutants: Oil. The main United Kingdom legislation dealing with pollution of the sea by shipborne oil is the Oil in Navigable Waters Acts, 1955, 1963 and 1971, as consolidated in the Prevention of Oil Pollution Act, 1971. The original purpose of the 1955 and 1963 Acts was to give effect to the International Convention for the Prevention of Pollution of the Sea by Oil, 1954, as amended in 1962 [12]. The 1971 Act amended the previous legislation in order to enable Her Majesty's Government to accept and bring into force various amendments to the Convention adopted by the Assembly of the Inter-Governmental Maritime Consultative Organisation (IMCO) in October 1969 [13]. The 1971 Act also introduced various measures for strengthening the control of oil pollution. These include, in particular, the increase in the penalties for various offences in connection with the prevention of oil pollution, as well as the granting to the government of powers of intervention to deal with shipping casualties which cause or threaten pollution by oil. All these Acts are now consolidated into the Prevention of Oil Pollution Act, 1971 [14].

(a) This prohibits the discharge of oil or oily mixtures (as defined) from *any* vessel into United Kingdom waters, i.e. into any part of the sea within the seaward limits of the territorial waters of the United Kingdom and all other waters including waters which are within these limits and are navigable by sea-going ships [15]. This provision applies to *all* ships irrespective of the state of registration and nationality. The prohibition also applies to waters in harbours.

(b) The Act also prohibits the discharge of oil and oily mixtures from any ship registered in the United Kingdom into any part of the sea outside the territorial waters of the United Kingdom [16].

The Act however empowers the Minister (the Secretary of State for Trade and Industry) to exempt any vessels or classes of vessels from any of the provisions of the Acts or any regulations made thereunder either absolutely or subject to such conditions as he thinks fit [17].

The 1955 Act had a provision prohibiting the discharge of certain oils into certain sea areas, i.e. into areas designated as 'prohibited zones'. The 1971 Act amends this provision and, following the principle of 'total prohibition' contained in the amendment adopted by the IMCO Assembly, now prohibits the discharge of any oil to which the Act applies, i.e. oil or oily mixtures (as defined in the Act) into 'any part of the sea outside the territorial waters of the United Kingdom [18]'.

The effect of this new provision is that all discharge of oil is prohibited until and unless the Minister has permitted any such discharge by virtue of the powers conferred on him by the original Act.

As in the Oil in Navigable Waters Acts, the 1971 Act empowers the Secretary of State for Trade and Industry, for the purposes of preventing or reducing discharges of oil and oily mixtures into the sea:

1. To make regulations requiring British ships registered in the United Kingdom 'to be fitted with such equipment and comply with such other requirements as may be prescribed ... and approved from time to time by persons appointed by him ...'.
2. To make regulations requiring masters of British ships registered in the United Kingdom to keep records of occasions in which oil has been discharged or escaped from ships, in particular discharges for the purpose of securing the safety of the ship or the safety of any vessel or of preventing damage to any vessel or cargo as well as the carrying out of operations relating to disposal of oil residues.
3. To make regulations requiring *all* vessels while they are within the seaward limits of the territorial waters of the United Kingdom, to keep records relating to the transfer of oil to and from other vessels and to require certain acts in relation to the use of the records.
4. To appoint a person to inspect and report in compliance with the prohibitions, restrictions and obligations imposed by virtue of the Act. There are penalties for the non-observance of regulations made under the Acts [19].

The Act prohibits the transfer of oil from or to vessels at night in any harbours (unless prescribed notice has been given to the harbour master). There are powers to harbour authorities to provide facilities for the reception of oil residues and the Secretary of State may require the provision of these facilities where none exist or where they are inadequate. There are various obligations on harbour authorities, shipmasters and occupiers of land to report incidents involving the discharge or escape of oil.

The Act empowers the Secretary of State, by Order-in-Council, to apply to non-British ships, while they are in United Kingdom harbours or territorial waters, regulations applicable to British registered ships relating to the fitting or equipment and the compliance with other requirements under the Act, as well as the keeping of records. There is to be no discrimination against countries but power is granted to the authorities to accept foreign requirements which are essentially similar to United Kingdom requirements and to exempt from the provision of the section ships which comply with such (acceptable) foreign requirements. However, the regulations are not applicable to ships which happen to be within United Kingdom waters solely by virtue of distress, i.e. stress of weather or any other circumstances which neither the master nor the owner (or charterer) of the ship could have prevented or forestalled [20].

Similarly, the Secretary of State, by Order-in-Council, is empowered to designate persons to go on board any ship to which the 1954/62 Oil Pollution Convention applies while such a ship is within a United Kingdom harbour and to require the production of any records required to be kept in accordance with the Convention or with any subsequent Convention so far as it relates to the prevention of pollution of the sea by oil [21].

The Act applies to all of the United Kingdom, including Northern Ireland, and provides for the extension to the Isle of Man, the Channel Islands or any colony or other country in which Her Majesty's Government has, for the time being, jurisdiction. The Convention's provisions, while not applicable to vessels of Her Majesty's Navy nor to government ships in the services of the Secretary of State while employed for the purposes of Her Majesty's Navy, apply nevertheless to government ships registered in the United Kingdom and to government ships not registered which are held for the purposes of Her Majesty's Government in the United Kingdom.

The Act contains special defences against the prohibitions contained therein. It is a defence against a charge under section 1 of the Act for a master or owner of a vessel to prove that the oil was discharged for the purpose of securing the safety of any vessel or of preventing damage to any vessel or cargo or of saving life. Such discharge must however be reasonable and necessary in the circumstances [22].

Where the oil escaped or leaked from the vessel or place, it will be a defence that neither the escape (leakage) nor any delay in discovering it was due to any want of reasonable care and that as soon as escape or leakage was discovered all reasonable steps were taken to reduce ... the effect. Similar defence applies to escape of oil due to damage of the vessel. In respect of discharge of oil contained in an effluent produced by operations for the refining of oil it is a defence that it was not reasonably practicable to dispose of the effluent otherwise than by discharging it into the water ... and all reasonable measures had been taken to eliminate oil from the effluent [23].

The 1971 Act increases the penalties for various contraventions from £1,000 to £50,000 [24]. In addition the 1971 Act, under section 12, confers various powers on the Secretary of State for Trade and Industry to take measures in cases of shipping casualties which in his opinion will or may cause pollution on a large scale in the United Kingdom or in the waters in or adjacent to the United Kingdom up to the seaward limits of the territorial waters. These measures include:

(a) The giving of orders to the owner of the ship or other persons capable of dealing with it in order to require certain actions taken or to require that they refrain from taking certain actions.

(b) Where such orders are or prove to be inadequate in the circumstances take further measures, including the sinking or destruction of the ship.

Section 14 of the Act provides penalties for failure to comply with the orders, or wilful obstruction, of any person who is acting in compliance with directions under the Act. There are provisions requiring endeavours to avoid risk to human life and also providing a defence to persons who fail to comply with directions which would have involved a serious risk to human life. The provisions of this section apply primarily to ships within the United Kingdom jurisdiction, i.e. ships registered in the United Kingdom and to other ships only so long as they are within the territorial waters of the United Kingdom. However, section 16 provides that the section may, by Order-in-Council, be made to apply to non-British registered ships outside British territorial waters if it is necessary to do so. The Act further states, in section 12(7) that the provisions of the section are without prejudice to any rights or powers of Her Majesty's Government in the United Kingdom exercisable apart from this section whether under international law or otherwise.

As stated by the Government 'the purpose of the section is to provide the Government with powers in respect of such ships comparable to those which are considered to be available under international law in respect of ships of other nationalities on the high seas, and which are set out in the International Convention Relating to Intervention on the High Seas in Cases of Oil Pollution Casualties, 1969 [25]. Section 8(10) of the 1971 Oil in Navigable Waters Act (which is now section 16 of the 1971 Prevention of Oil Pollution Act) enables appropriate parts of the Act to be extended by Order-in-Council to ships outside United Kingdom jurisdiction. The purpose of this is to remove any possible doubt about the powers of Her Majesty's Government in the United Kingdom domestic law to fulfil their obligations under international law, in particular the 1969 Convention. There is no intention to take action going beyond that permitted by international law. Her Majesty's Government have introduced this legislation in advance of the entry into force of the Convention in view of the urgency of the situation in the congested waters of the English Channel, and on the assumption that the Convention will enter into force: it remains their hope that the Convention will be brought into force as soon as possible' [26].

The Act provides, in section 13, that a person incurring expense or suffering damage as a result of action taken by himself or by others in pursuance of directions given under the appropriate Act shall be entitled to recover compensation from the Secretary of State if the action taken was not reasonably necessary to deal with the danger or if the damage done was not proportionate to the danger it was intended to cover. The section also confers jurisdiction on the United Kingdom courts to hear and determine claims arising under the Act.

It can be seen, therefore, that the United Kingdom legislation relating to pollution of the sea by shipborne oil implements, in essence, the provisions of the two major international conventions which are aimed at preventing oil pollution from ships: the 1955 and 1963 Oil in Navigable Waters Acts implement the 1954/62 Convention dealing with deliberate discharges of oil from ships (and other routine discharges or escapes arising from the normal operation of ships), while the 1971 Oil in Navigable Waters Act implemented the 1969 amendments to this Convention and goes further to implement the principles contained in the 1969 Convention which is designed to prevent oil pollution arising from maritime casualties. These provisions have now been brought together in one Consolidating Act which implements the two Conventions, and, in some cases, go beyond the strict requirements of either Convention.

Pollution from other shipborne pollutants.   Apart from legislation designed to prevent the discharge of noxious and polluting substances into the rivers, estuaries, tidal waters and into United Kingdom territorial waters (including legislation for the purpose of conserving fishery resources and the arrangements for regulating the dumping of industrial and domestic wastes into international waters from the United Kingdom mainland) there is no systematic United Kingdom legislation relating to pollution of the high seas by other shipborne substances. The Royal Commission on Environmental Pollution has recommended that the Government take action in this area [27].

As far as the discharge of rubbish and sewage from ships (i.e. rubbish and sewage arising from the operation of the ship), there are no statutory controls although Notices to Mariners are issued by the Department of Trade and Industry to draw attention to the dangers arising from the dumping of certain objects such as oil drums, lengths of wire, etc. With regard to any pollution which may arise from the handling of cargo other than oil, there are again controls through legislation and subsidiary legislation as far as rivers, harbours and coastal waters are concerned but no statutory controls outside territorial water. It is, however, admitted that pollution can arise from these operations particularly as a result of the jettisoning of soiled and contaminated dunnage,* cargo mats and clothes, and the deck washings and hold sweepings arising from a number of different cargoes such as grain, ores, chemicals, sulphurs, etc. [28].

Pollution arising from the exploration and exploitation of sea-bed resources.   The current United Kingdom legislation in this area is contained in section 3 of the Prevention of Oil Pollution Act, 1971. This replaces the provisions contained in Section 5 of the Continental Shelf Act of 1964 which was enacted to implement the provisions of Article 5(7) of the Convention on the Continental

* Mats, brushwood stowed with cargo to prevent moisture and chafing.

Shelf, 1958 [29]. Under this Convention, contracting states are obliged to take appropriate measures for the protection of the resources of the sea from harmful effects. Under the 1971 Act, if oil or mixture containing oil is discharged or escapes into any part of the sea (i.e. whether within or outside UK territorial waters) from a pipeline (or otherwise than from a ship) as a result of any operations for the exploration of the sea-bed and subsoil or the exploitation of their natural resources in a designated area the owner of the pipeline or the person carrying out the operation is guilty of an offence [30]. It is however a defence if such person is able to prove in the case of a discharge from a place in his occupation, that the discharge was due to the act of a person who was there without his permission (express or implied) [31]. It is also a defence for the person charged in respect of the escape of oil under the section to prove that neither the escape nor any delay in discovering it was due to any want of reasonable care and that as soon as practicable after it was discovered all reasonable steps were taken for stopping or reducing it [32].

In addition there are the Petroleum (Production) Regulations, 1966, which were made under the Continental Shelf Act of 1964. These regulations remain applicable both under the original legislation and under the 1971 Act. They provide for the issue of licences and the conditions to be attached to such licences for the exploration and production of petroleum products. Only persons (natural or body corporate) so licensed may engage in operations for exploration and production of these resources.

The regulations contain instructions on safe practice in drilling and production operations to which licensees are required to adhere strictly. For example, a licensee is required, under Clause 16 of the Schedule 4 (Model Clauses for Production Licences in Sea-Ward Areas) to maintain all apparatus and appliances and all wells ... in good repair and condition ... (and to) execute all operations in or in connection with the licensed area in proper and workmanlike manner in accordance with methods and practice customarily used in good oil field practice [33]. In particular the licensee should take all practicable steps to prevent the escape of petroleum into any waters in or in the vicinity of the licensed area [34]. Similarly the licensee is required to comply with any instructions from time to time given by the Minister in writing and relating to any of the matters dealt with under the appropriate paragraphs of the regulations [35]. The licensee is further required to give notice to the Minister of any event causing, *inter alia*, escape or waste of petroleum [36]; and the licensee shall not carry out any operations authorised by his licence in such manner as to interfere unjustifiably with, *inter alia*, the conservation of the living resources of the sea [37].

Persons authorised by the Minister are entitled at all reasonable times to enter the licensee's installations and equipment and to inspect or make abstracts of the records which the licensee is required to keep under the provisions of his licence [38]. And a licence to operate may be revoked for, among others, a breach or non-observance by the licensee of any of the terms and conditions of the licence [39].

While these regulations and the instructions and conditions contained therein cover a very comprehensive area, they obviously cannot deal with all the multifarious aspects of an activity which is developing so rapidly in terms of areas covered and practices involved. To supplement the provisions of the relevant legislation the operators, organised by the Institute of Petroleum, have developed recommended standards and practices designed to help in devising pollution controls and to give guidance to all operators in fulfilling their legal obligations as contained in the various enactments and subsidiary legislation in the field. This has been done by the drawing up of Codes of Practice for Exploration and Production. Operators are required to ensure that their drilling and production operations are governed by strict adherence to the safety and other instructions contained especially in Schedule 4 of the Petroleum (Production) Regulations of 1969. To all matters not specifically covered by these instructions, the provisions in the current edition of the Institute of Petroleum Code of Safe Practice for Drilling, Production and Pipeline Operation in Marine Areas will apply [40].

The instructions in this Code of Practice cover a whole range of subjects including the safety, health and welfare of the persons engaged in the operation as well as the safeguarding of other activities in the area of the operations. Great emphasis is naturally paid to the prevention of pollution. For example, in the very first chapter the operator is warned about the dangers of pollution and given instructions for dealing with pollution incidents.

'Oil spills constitute . . . a threat to marine life and cause nuisance on foreshores and beaches. They should be contained immediately they occur and prompt action should be taken to remove them either chemically or physically. No crude oil, waste oil, oil sludge or water emulsion or oil-bearing mixtures should be discharged or allowed to flow into any stream, lake or open sea' [41].

Then there are the more technical instructions relating to equipment and procedures actually employed in the operations. There are also detailed instructions on drilling (installation of well-head control equipment and pressure testing, etc.), production and drainage. There are detailed procedures also for avoiding 'blowouts' described as the source of the 'most serious threats of pollution in these areas' [42]. In addition there are instructions on the design, routeing, construction, laying, maintenance and inspection of sub-marine pipelines.

These instructions are strictly adhered to and, for the oil industry, constitute a very important part of the arrangements for avoiding and dealing with oil pollution in connection with exploration and exploitation of the sea-bed [43].

There is no United Kingdom legislation as yet on pollution from substances other than oil which may arise as a result of operations for the exploration and exploitation of the sea-bed and subsoil and the resources thereof.

Pollution from radio-active substances. The main United Kingdom legislation dealing with pollution of the sea from radio-active substances is contained in the Radio-active Substances Act of 1960. Under this, discharges of radio-active wastes, mainly from nuclear-powered electricity generating stations and Atomic Energy Authority establishments, into the sea and estuaries are controlled. The Act contains prohibitions against discharges of radio-active wastes except with the authorisation of appropriate authorities [44]. In addition, section 7 of the Continental Shelf Act of 1964 provides that regulations issued under the Radio-active Substances Act may be made applicable, by order under section 3 of the 1964 Act, to installations used for the exploration of the sea-bed and subsoil and the exploitation of the resources thereof.

Administrative Arrangements [45]

In the United Kingdom the administrative arrangements for dealing with pollution of the sea involve the participation and functions of a large number of departments and agencies: central, regional and local. Responsibility for the control of pollution of the sea from land-based pollutants (including radio-active wastes) is shared between the Department of the Environment which incorporates, *inter alia*, the Ministry of Housing and Local Government and the Scottish and Welsh offices, on the one hand, and the various river authorities (including the authorities responsible for the Thames, Lee and London areas) and harbour authorities on the other. In addition, the Ministry of Agriculture, Fisheries and Food, and the Department of Agriculture and Fisheries for Scotland, play a significant part in the programmes for the control of sewage pollution as well as the discharge of industrial, domestic and agricultural wastes. Other agencies and authorities participate in the arrangements. Local Fisheries Committees, the Central Electricity Generating Board and to some extent the Department of Trade, all participate at different levels in the programme designed to control and regulate the effects of pollution from land-based pollutants. The Ministry of Power is responsible for the control of pollution from radio-active wastes. The Foreign and Commonwealth Office is responsible for the foreign policy aspects of environmental issues, including the conclusion of international agreements for the control of marine pollution.

With regard to the control of oil pollution, the administrative arrangement is even more complicated. Responsibility is divided among several agencies. The main responsibility for operating the Prevention of Oil Pollution Act, 1971 lies ultimately with the Department of Trade and Industry although the Secretary of State may give various authorisations to a number of local agencies such as the harbour authorities, including in particular the Port of London Authority. The various regulations and orders required or permitted to be made are all the responsibility of the Department. When, however, it comes to dealing with actual incidents of oil pollution then there is a more complex division of labour. Responsibility is then divided between local authorities and a number of departments of the central government. The arrangements vary according to whether the polluting oil is (a) on the beaches or floating up to about one mile off-shore; or (b) floating more than one mile off-shore. Where oil is on beaches or floating up to one mile off-shore, the primary responsibility is that of local authorities whose areas include parts of the coast. These authorities have been asked by the central government (the Ministry of Housing and Local Government and the Scottish and Welsh Offices) to draw up schemes for dealing with such cases. Such schemes are always drawn up in consultation with river authorities, dock and harbour authorities and county councils, as the case may be. These schemes include appointment of officers to deal with pollution and the provision of manpower, technical advice and equipment. In serious cases, local authorities may call for the help of the armed services—usually the Navy and the coast guard, but they have to make payment for the services rendered.

The expenditure incurred in the fight against pollution is reimbursed in part from central government funds.

Where the polluting oil is more than one mile off-shore, the responsibility falls on the central government through the Department of Trade and Industry. The Department exercises this responsibility through the coast guard service and the principal officers of marine survey districts. There are nine such districts covering the whole of the coastline of the United Kingdom. In particular the Department has established extensive reporting arrangements under which all United Kingdom ships and aircraft —civil or military—are asked to report any incidents or casualties, involving or likely to cause oil pollution. There is constant co-operation with other departments of the central government, chiefly the Ministry of Defence, the Ministry of Technology, the Ministry of Agriculture, Fisheries and Food and the Scottish and Welsh offices. In addition, consultation and collaboration goes on with quasi-governmental and non-governmental scientific establishments such as the Natural Environmental Research Council (Nature Conservancy) and with the shipping and oil industries. Responsibility for the control over exploration for an exploitation of oil and gas resources off-shore lies with the Ministry of Power.

The Department of the Environment brings together into a single unit three great ministries (the Ministries of Transport, Local Government and Housing and Public Building and Works) which, among them, cover in England and Wales the overwhelming bulk of national planning and pollution control. The Secretary of State for the Environment has responsibility for co-ordinating all the works done by the government to control pollution in all departments [46].

These arrangements provide on the whole a co-ordinated system for the prevention and control of pollution of the sea from various sources. It has, however, been suggested that the arrangement dividing responsibilities between the central government and local government authorities might be more streamlined. In particular it has been suggested that the areas of jurisdiction of coastal local authorities might be extended in a seaward direction to ensure that all relevant functions, including planning, relating to the prevention and control of pollution might be better co-ordinated [47]. In relation to the special problem of oil pollution it has been said that the division of responsibility between the Department of Trade and Industry and local government authorities (and river authorities) by reference to the distance from the coast where the oil is floating could cause confusion and render the control of oil pollution less effective. Under the present arrangements oil floating less than a mile from the coast falls within the jurisdiction of local authorities and oil floating beyond one mile from the coast falls within the jurisdiction of the Department. The suggestion has been made that this might create problems in special parts of the coast and that therefore 'a limit based on depth of water with variations if necessary to suit local conditions would be far more practical' [48].

Comparative Legislation in Other Countries

Pollution from land-based substances.   In terms of objectives and content, legislation in other countries follows the same general pattern as the United Kingdom legislation. In most countries the legislation dealing with the matter is oriented mainly—if not solely—to the objective of safeguarding the quality of inland waters and the protection of health. It is only incidentally that the legislation is directed to the protection of the marine environment as such. However a few countries have, in recent years, placed increased emphasis on the problem of the prevention of marine pollution. Some of these have now included legislation dealing specifically with the prevention of marine pollution from pollutants entering the sea from the land. In New Zealand and Norway there are provisions controlling activities on land which cause or may cause pollution of the sea [49]. In the United States the declared purpose of the marine legislative measure in relation to pollution control is stated to be 'to enhance the quality and value of water resources and establish a policy for the protection, control and abatement of water pollution'. The Act authorises the development of a comprehensive programme for, *inter alia*, 'eliminating and reducing the pollution of inter-state waters', which term includes coastal waters. Under the Act pollution of inter-state or navigable waters 'which endangers the health or welfare of any person' shall be subject to various measures of abatement 'whether the matter causing or contributing to such pollution is discharged directly into such waters or reaches such waters after discharge into a tributary of such waters' [50].

In Canada recent legislation includes provisions designed to control and regulate activities which are likely to cause pollution of the sea whether such activities take place on land or at sea [51]. The Act provides a system of controls based on the need to obtain prior authorisation before any waste of any kind is deposited into the applicable sea areas. Depositing without such prior authorisation is punishable. There is also strict civil liability for damage caused by such unauthorised deposits. The term waste is defined as 'any substance detrimental to use of the . . . waters by men or fish and to plants men use' and it includes 'detrimentally altered waters'. Any deposit of waste into the waters or on land is prohibited if it may enter (applicable waters) or lead to waste which enters them.

These enactments appear to portend a clearer recognition by states of the importance of including control of land-based pollutants in any overall strategy to control and prevent pollution of the seas.

Dumping of wastes into the sea. Here too the pattern of legislation and administration in most countries is similar to what we have in the United Kingdom. Generally there is national legislation (administered centrally or through local authorities) which grants various powers of control and prohibition in respect of the dumping of wastes into coastal waters or other specified areas of territorial seas. As in the United Kingdom there is generally very little formal statutory control of dumping of wastes into international waters. There are however a few states which control dumping of wastes into the sea, even beyond the limits of territorial waters. Thus Canada has, in the Water Act of 1970, extended state powers to control over dumping from barges beyond territorial limits. A new legislation in the Netherlands controls wastes dumping 'beyond the limits of Dutch territorial waters'. This control applies to Dutch ships but extends to ships of other nationalities where they dump wastes which have been transported through Dutch territory [52]. In the United States a new legislation has been introduced to deal with the problem [53]. This measure declares itself to be the policy of the United States to regulate the dumping of 'all types of materials in the oceans, coastal and other waters and to prevent or vigorously limit the dumping . . . of any material which could adversely affect human health, welfare or amenities of the marine environment, ecological systems or economic potentialities'. For this purpose the measure seeks to regulate the transportation of materials from the United States territory 'for dumping into the oceans, coastal waters and other waters'. It also seeks to regulate the 'dumping of materials by any person from any source if the dumping occurs in waters over which the United States has jurisdiction'.

In the Federal Republic of Germany although there is no legislation as such, a voluntary system of control exists in relation to the dumping of wastes into the high seas from the territory of the Federal Republic of Germany by nationals of the Republic [54].

Marine pollution from shipborne pollutants. As in the United Kingdom, the legal regime for controlling this type of pollution relates almost entirely to pollution by oil. And once again the legislation is in almost all cases based on the International Convention for the Prevention of Pollution of the Sea by Oil, 1954 as amended in 1962. Some national legislation also incorporates the amendments to the Convention adopted by the IMCO Assembly in 1969, even though these amendments are not yet in force [55]. Most legislation in this area is confined to the implementation of the Convention with its amendments. Thus the French legislation, the Law of 1964, deals only with the discharge of oil and oily waste from ships [56]. The Act contains penalties for infractions at sea or in territorial waters by French ships and applies penalties to foreign ships for infractions while in French waters. This applies to ships of all nationalities whether or not they belong to or are registered in states parties to the 1954/62 Convention. It also defines the method of procedure in court, designates the officials who may institute proceedings and provides for claims for damages to the public foreshores. The law is supported by two Decrees, one regarding the form of Oil Record Books and the other defining the Classes of Vessels exempt from the provisions of the law [57]. On the other hand, the United States legislation in this area goes much further than the 1954/62 Convention. Like the United Kingdom Oil in Navigable Waters Act, 1971, the United States legislation incorporates the provisions contained in the 1969 amendments to the International Convention for the Prevention of Pollution of the Sea by Oil, 1954–62, as well as the provisions of the 1969 IMCO Convention Relating to Intervention on the High Seas in Cases of Oil Pollution Casualties. Additionally the United States Legislation also incorporates provisions from the second 1969 IMCO Convention: the Convention on Civil Liability for Oil Pollution Damage [58]. The Act defines 'oil' as 'oil of any kind or in any form . . . and includes oil mixed with wastes other than dredged spoil'. The Act prohibits the discharge of oil into or upon the navigable waters of the United States, adjoining shorelines or upon the waters of the contiguous zone in harmful quantities as determined by the President by regulation [59]. However even the protection of the waters of the contiguous zone is limited to pollution 'which threatens the fishery resources . . . or threatens to pollute or contribute to the pollution of the territory or territories of the United States' [60]. The Act applies to the discharge or escape of oil from vessels and from off-shore facilities [61]. The Act grants powers to the government to take measures to require the removal of such oil and, if necessary, to cause such oil to be removed at the expense of the person responsible therefor [62]. In the case of a maritime disaster creating a substantial threat of pollution hazard, there is power to remove the vessel and, if necessary, destroy such vessel by whatever means available [63]. Further there is power to secure from the federal courts of the United States such relief as may be necessary to abate such threat [64].

The Act imposes civil penalties for breach of its provisions and imposes civil liability on the owner or operator of vessels or off-shore facility in respect of the cost of removing the damage. This liability is, in general, in respect of any single occurrence, although where the person charged is guilty of wilful negligence or wilful misconduct they will be liable for the full amount of the costs of removal of the oil [65]. There are various defences chief among which are that the discharge was caused solely by an act of God, act of war, negligence on the part of the United States government or an act of a third party.

To cover the various liabilities under the Act, the owner or operator of a vessel is required to establish and maintain evidence of financial responsibility to meet the maximum liability to which his vessel could be subjected under the Act [66]. Provision is also made for measures to be taken to provide financial responsibility for all on-shore and off-shore facilities [67].

Legislation of similar scope has also been passed in Canada [68].

Pollution from shipborne substances other than oil. There are very few countries which have legislation dealing with the control of pollution of international waters from cargoes other than oil. Where legislation exists in respect of such cargo, it deals mainly with pollution which occurs in coastal waters or other waters subject to jurisdiction. Only in a few cases, as in the United States and Canada, has legislation been enacted to deal with pollution of the sea from cargoes other than oil. The Canadian Act of 1970 defines waste as 'any substance which alters waters to an extent that it is detrimental to their use by man' [69]. In the United States a new legislation deals with hazardous substances other than oil [70]. These are defined as such elements and compounds which when discharged in any quantity into or upon navigable waters . . . present an imminent and substantial danger to public health or welfare, including but not limited to fish, shellfish, wildlife, shore-lines, beaches. The Act contains prohibitions, authorisations and penalties for discharge of such substances.

Pollution arising from the exploration of the sea-bed and subsoil and the exploitation of its resources. It is now generally recog-nised that the increasing activities related to the exploration and exploitation of the resources (chiefly mineral petroleum) of the sea-bed pose a considerable pollution hazard. For this reason many countries have introduced, or are introducing, legislation to control such activities with a view, *inter alia*, to ensuring that they are conducted with due regard to the need to prevent pollution of the super-adjacent waters. Although such operations usually take place outside the territorial limits of states, the coastal state is empowered to, and does, exercise legislative jurisdiction. Article 2 of the 1958 Convention on the Continental Shelf grants to the coastal state the power of exercising sovereign rights for the purpose of exploring it and exploiting its natural resources. Article 5(7) of the Convention enjoins states to take all appro-priate measures in the contiguous zone for the protection of the living resources of the sea. On the basis of these provisions, many states have enacted legislation to control activities of exploration and exploitation of the resources of the sea-bed of the waters con-tiguous to their territorial limits. Most of the legislation, as in the United Kingdom, is confined at present to pollution from oil.

Generally legislation prohibits pollution arising from activities of exploration and exploitation and attaches various sanctions and penalties to infractions of these prohibitions. In addition many countries follow the United Kingdom system whereby operations are subject to state licensing. Usually the licence will contain a set of conditions requiring operations under it to follow certain procedures to prevent or stop pollution if it occurs [71]. In some cases the conditions are set out in the licence while in others the conditions are contained in general regulations which are merely referred to in the specific licences. In the United States new legis-lation makes oil companies liable for the cost of removal of oil pollution from drilling rigs up to a maximum of $8 million [72].

Pollution of the sea by radio-active substances. Most countries with nuclear programmes do have legislation aimed at controlling pollution by radio-active substances. In almost all cases the legislation deals with the disposal of all radio-active substances; and it is very rare for legislation to deal directly with disposal of radio-active wastes into the sea. Indeed there are countries which, like the United Kingdom, have removed the control of radio-active pollution from the purview of the authorities which deal with other aspects of water pollution in general. But other countries include radio-active pollution of sea-waters within the general framework of the law dealing with water pollution [73].

International Regulation and Policy [74]

In spite of general dissatisfaction with what has been achieved so far by international action in the field of marine pollution control, it is perhaps fair to say that there are very few areas in which there has been so much recognition of the relevance and need for international action and, in which relatively speaking, there has been so much success—both in regard to concrete achievement and to hopeful promise. Nor is it surprising that this should be so. For, in a real sense, marine pollution is, if not a global problem requiring a global approach to its solution, at least a problem which in many areas involves more than one state. Very few pollution incidents or tendencies can be localised to single states or can be considered to be matters of exclusively local concern: many have international implications. And, as the recent controversy between Canada and the United States in relation to the former's control measures against pollution of the Arctic waters clearly demonstrated, actions or omissions of other states may have serious implications for the interests of other states [75]. For this reason, it has for a very long time been recognised that the best, though not necessarily the only, way of dealing with the problem of marine pollution is at the international level. And in many areas and in respect of various types of pollution, there now exists clearly identified international policy and, in many cases, international schemes for the control and regulation of pollution of the seas and oceans—both for areas within territorial waters of states and for areas outside those limits.

The various international regimes differ in comprehensiveness and effectiveness, depending on the type of pollution with which they are designed to deal. For some types of pollution there is hardly any regime at all—apart from the general declaration of policy that pollution is to be avoided. On the other hand there are fairly elaborate, even if not as yet completely effective, systems in relation to other types of pollution.

In regard to pollution of the sea arising from land-based activities, there has until very recently hardly been any serious attempts at international control. This is due at least in part to the fact that where pollutants are discharged from land-based activities, they usually are discharged directly into territorial waters and only indirectly into international waters. The idea has therefore gained ground that the control of such pollution is a matter within the exclusive jurisdiction of coastal states. Another reason why international action in this area has been generally lacking is, perhaps, the fact that 'wastes are disposed of in the oceans (and seas) because the costs and risks of putting them elsewhere are greater' [76]. Whatever the reason, the international community seems to have accepted the view that it is left to each state to regulate as best it may the pollution of the sea which arises from activities within its territory. At the global level, the only act remotely resembling an attempt to deal with this problem is contained in the 1958 Convention on the High Seas. Article 25 of the Convention provides in paragraph 2 that states shall 'cooperate with competent international organisations for the prevention of pollution of the seas or air space above resulting from activities with radio-active materials or other harmful agents'. It has been suggested that wastes dispersed from land might, if they are sufficiently toxic, fall within the category of 'other harmful substances' and, consequently, that states would be obliged to cooperate with international organisations in taking measures to prevent their disposal into the seas [77]. It has, however, been recognised that this obligation, such as it is, is preconditioned on the adoption by a competent international organisation of recommendations or regulations for measures. To date no such recommendations or regulations have been adopted in respect of pollution from land-based activities [78]. It has also been suggested that a means of international control exists in the sense that a state which suffers damage as a result of the discharge (or dumping) into the sea of pollutants from land-based activities might bring a claim on the basis of general international law if it could show that it has suffered injury within its territorial limits or elsewhere. But again it has been clearly recognised that the problem does not lend itself to adequate treatment through international claims and that 'what is needed is action by an international organisation [79]'. It may be added that what is required need not necessarily be action by an international organisation. What is needed is action properly co-ordinated at an appropriate international level. One such measure, at a regional level, is the European Convention on the Limitation of the Use of Certain Detergents in Washing Powder and Cleaning Products. This agreement, opened for signature in 1968 by the governments of eight European states, provides a useful pointer to the kind of action which can and must be taken in this area. It may well be that this type of pollution can best be controlled for the time being, at least, at the regional level.

In relation to marine pollution arising from the dumping of wastes into the sea, there is hardly any effective international control measure in force. Again, this may have been due partly to the fact that most states found it necessary and convenient to dump into the high seas wastes which could not be safely or economically dumped in internal waters and partly to the fact that the area of the high seas has traditionally been subject to the authority of no state. Despite the provision of the 1958 Convention on the High Seas, that states shall co-operate to prevent pollution of the sea from 'harmful agents', it has not been the policy of the international community to regulate in any way the disposal into the high seas of wastes and other substances by states and other persons. Even though the polluting potentialities of such substances have been recognised, no attempt has been made, until quite recently, to do anything about the problem created by their disposal.

However, in the past half-decade the situation has changed. Various schemes, at inter-state and regional levels, have been initiated to control the dumping of various harmful substances into the seas. An example of this is the negotiations currently under way among Scandinavian countries to restrict the dumping of chlorinated hydrocarbons and certain other heavy metals into the sea [80]. An even more significant development was the adoption, in February 1972, of the 'Convention for the Prevention of Marine Pollution by Dumping from Ships and Aircraft' by the governments of twelve European states. This Convention, held in Oslo (Norway), is designed to protect the waters of the North East Atlantic Ocean, including North Sea [81]. The contracting parties agree to take all measures to prevent the pollution of the sea by dumping and to harmonise their policies and introduce individually and in common, measures to prevent such pollution [82]. Although this is a regional Convention designed to apply to a well-defined area, the contracting parties also agree to apply the measures which they adopt (for their region) in such a way as to prevent the diversion of dumping of harmful substances into seas outside the area to which the Convention applies [83].

At the global level considerable work was done before and at the United Nations Conference on the Human Environment (Stockholm, June 1972) on the problem of ocean dumping, within the general context of the preservation of the marine environment.

The Conference discussed a draft convention on the subject of ocean dumping which had been prepared by an inter-governmental meeting in Reykjavik (Iceland) in April 1972 and referred the draft for further international action with a view to opening the proposed convention for signature preferably before the end of 1972 [84]. On the general question of ocean dumping, the Conference recommended to governments to ensure that ocean dumping by their nationals anywhere, or by any person in areas under their jurisdiction, is controlled and that governments shall 'continue to work towards the completion of, and bringing into force as soon as possible of, an overall instrument for the control of ocean dumping as well as needed regional agreements within the framework of this instrument, in particular for enclosed and semi-enclosed seas, which are more at risk from pollution' [85].

These various attempts indicate a welcome awareness on the part of the international community that something needs to be done about the problem. While it is not yet certain that a really effective system will be forthcoming in the near future it is at least reassuring that the question has at last attracted the full international concern and discussion which it undoubtedly deserves.

Pollution from shipborne pollutants. As at the national level, action at the international level has related primarily, if not entirely, to the control and prevention of pollution from oil. A variety of legal measures have been taken at regional and global level to deal with the problem of oil pollution. A series of international treaties and instruments have been drawn up to regulate and prevent oil pollution caused either in the course of a ship's operation (i.e. pollution arising from routine oil spills, whether deliberate or negligent) or as a result of major accidents involving ships. The 1958 Convention on the High Seas provided in Article 24 that 'states should draw up regulations to prevent pollution of the seas by the discharge of oil from ships . . . taking into account existing treaty provision on the subject'. Already, in 1954 the International Convention for the Prevention of Pollution of the Sea by Oil had been concluded. This Convention, with amendments adopted in 1962, is designed to prevent the discharge of oil or oily mixtures in certain 'prohibited zones' and provides a means of co-ordinating the actions of states to make the prohibition effective. In 1969 and 1971 more amendments were adopted designed to extend further the scope of application of the Convention and to make even more effective the means of enforcement. In particular these amendments would do away with the notion of prohibited zones and, in effect, prohibit the discharge of oil into any areas of the sea unless under stringent well-defined conditions [86].

The enforcement procedures involve requirements for the maintenance of oil record books by ships, the fitting of special equipment to prevent accidental discharge or leakage of oil and the specification of special procedures for operations involving the receiving and discharge of oil. A state party to the Convention is authorised to take measures to find out whether these requirements have been met by ships of other contracting states, when such ships are within its ports. However there is no effective means of detecting violations of the Convention at sea and, even when violations have been detected, there is no effective system for ensuring that adequate (or any) penalties are in fact exacted against the offenders. It is entirely left to the flag state to determine what action and what penalties (if any) it will or can take against such offenders. These and other weaknesses have led to calls to amend the conventions and tighten its enforcement procedure. While there is justification for these calls, it should perhaps be said that the difficulties with the Convention do arise not so much from the substance and scope of the prohibitions as from the machinery of enforcement. And any strengthening of the Convention 'is likely to hinge primarily on the provisions for supervision of compliance rather than in the substance of the prohibition' [87]. This fact has been clearly recognised by governments and one of the measures scheduled to be considered at the conference being convened by IMCO in 1972 on the subject of marine pollution will relate to improved enforcement procedures [88].

The 1954 Convention with its various amendments deal only with the prevention of pollution from routine discharges of oil. Until 1969 there was no international instrument dealing with the problem of massive oil pollution arising out of maritime disasters. This problem was dramatically brought to the attention of the international community by the *Torrey Canyon* disaster of 1967. Following this accident, international action was begun within the consultative machinery provided by the Inter-Governmental Maritime Consultative Organisation (IMCO) which is also responsible for the administration of the 1954/62 Convention and its amendments. The discussions within IMCO led to the adoption in 1969 of the International Convention relating to Intervention on the High Seas in Cases of Oil Pollution Casualties [89]. This Convention aims at protecting the interests of peoples against grave consequences of a maritime casualty resulting in danger of oil pollution of the sea and coastlines by ensuring that in these circumstances necessary measures of an exceptional character to protect such interests can be taken. The Convention gives to a state party to it, the right to take 'such measures on the high seas as may be necessary to prevent, mitigate or eliminate grave imminent danger to (its) coastline or related interests from pollution or threat of pollution of the sea by oil, following upon a maritime casualty or acts relating to such casualty, which may reasonably be expected to result in major harmful consequences' [90]. There are well-defined conditions attached to the exercise of this right and a state which takes any measures in violation of these conditions or which takes action which is unreasonable or disproportionate in the light of the threatened danger, may be required to pay compensation for any damage resulting from such action.

This Convention, therefore, does in some way deal with pollution from accidents. However it is to be noted that even this Convention is restricted in its application. A contracting party can only take action if the pollution or threat of pollution is to 'its coastlines and related interests'. It cannot take action to prevent pollution of the seas, as such. Again, the right to take action is restricted to ships of contracting states. Other ships are protected by the principle that ships on the high seas are within the exclusive jurisdiction of the state whose flag they fly.

At the regional level there is the Agreement of Co-operation in Dealing with Pollution in the North Sea by Oil which was concluded in 1969 by eight countries bordering the North Sea [91]. This Agreement divides the North Sea into administrative zones for action on oil pollution threats and establishes a 'scheme of mutual co-operation and assistance to deal with oil pollution incidents'.

Pollution of the sea from shipborne pollutants other than oil. Owing mainly to the absence of adequate technical information particularly as to the various types of pollutants and their polluting potentialities, there has to date been no international legal measures to deal with pollution of the sea by agents other than oil. The 1969 IMCO Diplomatic Conference which adopted the Convention Relating to Intervention on the High Seas in Cases of Oil Pollution Casualties, noted this and while stating that the limitation of the Convention to oil was

'not intended to abridge any right of a coastal state to protect itself against pollution by any other agent, recommended IMCO to intensify its work in collaboration with other interested organisations in all aspects of pollution by agents other than oil and further recommended that contracting governments which become involved in a case of pollution danger by agents other than oil co-operate as appropriate by applying wholly or partially the provisions of this Convention' [92].

Following the work of the Group of Experts on the Scientific Aspects of Marine Pollution (a Group established and maintained by co-operation between various United Nations Specialised Agencies) [93], it is now felt that enough information has been collected and analysed to warrant an attempt at legal regulation on the lines followed in relation to pollution by oil. The proposed 1973 Convention of IMCO will deal not only with pollution by oil but also with pollution arising from deliberate and accidental spills of oil as well as other 'noxious and hazardous cargoes and substances'.

Pollution arising from the exploration and exploitation of the sea-bed and subsoil. The international regulation of this type of pollution finds its origins in the 1958 Convention on the Continental Shelf. Article 5, paragraph 1 of the Convention provides that the exploration of the continental shelf and the exploitation of its natural resources must not 'result in any unjustifiable interference with navigation, fishing or the conservation of the long resources of the sea'. Similarly, paragraph 7 of the same Article provides that a coastal state is obliged to take all appropriate measures for the protection of the living resources of the sea from harmful agents. However, until quite recently, the application of these general provisions were left entirely to state initiative and discretion. But over the past half-decade, increased concern of the international community with the problem of marine pollution from activities in the sea-bed area has led to a number of measures and proposals which can reasonably be expected to lead to some form of relatively effective international action. For instance, the UN Declaration of the Principles Governing the Sea-Bed and Ocean Floor, etc. contains a provision requiring states, in their activities in the area, 'to take measures and co-operate in the adoption and implementation of international rules, standards and procedures for *inter alia* prevention of pollution and contamination and other hazards of the marine environment including the coastline, and of interference with the ecological balance of the marine environment' [94].

The UN Conference on the Human Environment devoted a

substantial part of its work to the problem of marine pollution and produced a number of declarations and recommendations for future action on the subject. For example, Principle 7 of the 'Declaration' of the Conference holds that 'States shall take all possible steps to prevent pollution of the seas by substances that are liable to create hazards to human health, to harm the living resources and marine life, to damage amenities or to interfere with other legitimate uses of the sea' [95]. And the proposed UN Conference on the Law of the Sea is expected to devote considerable time and effort to the questions of marine environment. Indeed one of the sub-committees of the UN Committee charged with preparing for this conference [96] has as its terms of reference the preservation of the marine environment (including, in particular, the prevention of pollution) and scientific research.

Pollution of the sea from radio-active substances. The origins of international action relating to prevention of pollution of the sea by radio-active substances can be traced back to the 1958 Convention on the High Seas [97]. Article 25 of the Convention provides that:

1. Every state shall take measures to prevent pollution of the seas from the dumping of radio-active waste, taking into account any standards and regulations which may be formulated by the competent international organisations.
2. All states shall co-operate with the competent international organisations in taking measures for the prevention of pollution of the seas or air space above resulting from any activities with radio-active materials (or other harmful agents).

At the 1958 Conference a move was in fact made to adopt a provision banning all discharges of radio-active substances into the sea, but this proved unsuccessful.

The 1958 Conference adopted a resolution recommending that the IAEA should assist states by promulgating standards and drawing up internationally acceptable regulations relating to the discharge of radio-active materials into the sea [98]. Owing, however, to a number of factors, including political and scientific factors, it has not been possible for the IAEA to take any regulatory action beyond a series of procedural recommendations for monitoring and reporting [99]. These provide an international guidance system for controlled disposal of radio-active waste. At the regional level much more seems to have been achieved. Special mention might be made of the system set up by EURATOM countries in 1959 [100]. This system establishes basic standards including maximum permissible levels of contamination. States parties to the scheme are required to submit detailed information on any plans for disposal of radio-active waste at least six months before the plan is put into operation so that the EURATOM Commission may determine whether the operation is safe. Mention should also be made of the system whereby the disposal of radio-active substances into the Atlantic by a group of European States is organised and co-ordinated by the Nuclear Energy Agency of the OECD [101].

International action of another kind was taken in the 1963 Nuclear Test Ban Treaty which places a general prohibition on the testing of nuclear weapons, although the treaty contains an exception with respect to testing conducted beneath the sea-bed, provided it can be done 'without effects on the super-adjacent floor or water' [102]. In view of the fact that as stated by the IAEA, nuclear weapon tests form by far the largest cause of radio-activity in the sea [103], this Treaty, especially if and when it is accepted by all nuclear states, would go a long way to prevent pollution of the sea by radio-activity. In this connection also the proposed treaty on the prohibition of the emplacement of nuclear weapons and other weapons of mass destruction on the sea-bed and ocean floor and in the subsoil thereof which was presented jointly by the United States and USSR Governments at the Geneva Conference of the Committee on Disarmament, would go a long way towards diminishing 'the threat of pollution of the marine environment' [104].

Other regimes. In addition to international measures designed to prevent pollution of the sea through direct prohibition with or without any or adequate sanctions there are measures which attempt to prevent pollution by an indirect method. These are the measures which establish regimes of liability for acts which cause pollution where such pollution causes damage to third parties. On the hypothesis (which is generally assumed to be valid) that by making a party liable for damage caused by pollution (and especially by making his liability absolute or subject to as few defences as necessary) it is possible to get him to take extra care to avoid acts or omissions which are likely to cause pollution. Since most activities which cause pollution are engaged in for profit, it is felt that by making the penalty for pollution high enough, an economic deterrent would be created and persons engaged in potentially pollution-creating activities would take more care.

Partly on the basis of this principle a number of international agreements have established various regimes of liability for damage caused by pollution of the sea. Of these the following are the principal examples.

1. The 1969 Convention on Civil Liability for Oil Pollution Damage. This Convention provides for a system of more or less absolute liability on the shipowner for pollution damage caused by the discharge or escape of oil from his ship. A shipowner may be required to pay compensation up to a maximum of $14 million or, where he is guilty of fault, up to the actual cost of repairing the damage [105].

2. The Brussels Convention of 1962 on the Liability of the Operators of Nuclear Ships [106].

In 1971 a supplementary Convention to the 1969 Civil Liability Convention was adopted in Brussels by a conference convened again by IMCO. This Convention—the International Convention for the Establishment of an International Fund for Compensation for Oil Pollution Damage—established an international fund for which compensation will be paid to states and other persons which suffer pollution damage arising from marine casualties [107].

Prospects for Future Action

Marine pollution has now become an international issue demanding serious international attention. In the United Nations system, six organisations namely the FAO, IAEA, IMCO, UNESCO, WHO and WMO, have shown and continue to show direct interest in the problem of marine pollution and its control. A large number of states have indicated their concern with the problem and suggested that existing international arrangements should be fully implemented and perfected where necessary. As a result of the great international discussion and consideration of the problem the prospects for international action have become brighter over the past few years. In addition to various national and regional measures a number of international measures and arrangements are envisaged in the immediately foreseeable future. These include:

1. The 1973 UN Conference on the Law of the Sea. This Conference will deal *inter alia* with the protection of the marine environment from pollution as envisaged in the 1970 UN Declaration of Principles on the Sea-Bed area.

2. In 1973 IMCO is convening an international conference for the purpose of preparing a suitable international agreement for placing restraints on the contamination of the sea, land

and air by ships, vessels and other equipment operating in the marine environment.

The General Assembly of the United Nations indicated in its Resolution 2566 (XXIV) of 13 December 1969 that the United Nations might provide more comprehensive action and improved co-ordination in the field of the prevention of pollution of the marine environment. The Conference of 1972 and 1973 may therefore be expected to lead to the establishment of such a system to co-ordinate the efforts of various states and international organisations for the control of marine pollution and, ultimately, for the control of pollution in the whole of the human environment. It can only be hoped that the increased awareness, by states and commercial and industrial concerns, of the dangers of pollution will be able to sway the members of the United Nations to take appropriate and realistic action—before the situation gets out of hand.

References

1. See R. Toms, 'The prevention of river pollution', *Journal of Environmental Planning*, Vol. 1, No. 1 (1972), p. 37.
2. Rivers (Control of Pollution) Act, 1951, s. 2.
3. Ibid., s. 7.
4. Clean Rivers (Estuaries and Tidal Waters) Act, 1960, s. 11.
5. Water Resources Act, 1963, ss. 113–15.
6. Royal Commission on Environmental Pollution: Third Report 1972 (Cmnd 5054), p. 7, para. 19.
7. Ibid., pp. 9–12 and Chapter V.
8. See p. 186.
9. See United Kingdom replies to IMCO Inquiry contained in IMCO Document OP VIII/6(c) of 30 June 1970. Also *Human Environment— The British View* (HMSO, 1972), p. 34.
10. Royal Commission on Environmental Pollution. Third Report, op. cit., p. 10, para. 27(e).
11. See note 9 above.
12. United Nations Treaty Series, Vol. 600, United Kingdom Treaty No. 36 (1962), Cmnd 1711, p. 336.
13. By Resolution A. 175 (VI)—Text in *International Legal Materials* (1 January 1970).

14. This Act replaces s. 5 of the Continental Shelf Act, 1964, which related to pollution arising from the exploration and exploitation of the sea-bed and sub-soil.
15. Prevention of Oil Pollution Act, 1971, s. 2(2).
16. Ibid., s. 1.
17. Section 23, following the provisions of Article 11 of the Convention.
18. Prevention of Oil Pollution Act, 1971, s. 1.
19. Oil in Navigable Waters Act, 1955, s. 7 as amended by s. 4 of the 1971 Act; now s. 17 of the Prevention of Oil Pollution Act, 1971.
20. Ibid., s. 22(3).
21. Ibid., s. 21.
22. Ibid., s. 5(1).
23. Ibid., s. 6(2).
24. Ibid., ss. 1(4) and 2(4).
25. Concluded at Brussels on 29 November 1969, IMCO, London No. IMCO 1970:3. 9, *International Legal Materials*, p. 25.
26. See United Kingdom Note reproduced in IMCO Document OPC 2/ Circ. 13 of 1 June 1971.
27. Third Report, op. cit., pp. 13–14.
28. See IMCO Document OP VIII/6(c) of 30 June 1970.
29. United Nations Treaty Series, Vol. 499, p. 312.
30. Prevention of Oil Pollution Act, 1971, s. 3(1).
31. Ibid., s. 3(1).
32. Ibid., s. 6.
33. Petroleum (Production) Regulations 1966, Schedule 4, s. 16.
34. Ibid., s. 16(1)(e).
35. Ibid., s. 16(2).
36. Ibid., s. 16(3).
37. Ibid., s. 17.
38. Ibid., ss. 26–7.
39. Ibid., s. 33.
40. The Current Edition is the 1964 Edition which is Part VIII of the Institute of Petroleum Model Code of Safe Practice in the Petroleum Industry. See on this J. H. Edwards in *Journal of Environmental Planning*, Vol. 1, No. 1 (1972).
41. Code: op. cit., Chapter 1, para. 1.9.
42. Ibid., p. 13.
43. See J. H. Edwards, op. cit.
44. Radio-active Substances Act, 1960, s. 6. The Minister of Housing and Local Government is empowered to issue the authorisations. However, s. 8(2) of the Act requires him, before issuing any authorisations to consult a number of authorities including central government ministries and local and river authorities, as may be appropriate.
45. See United Kingdom Note on National Arrangements for Dealing with Significant Spillages of Oil issued in IMCO Document OPS/Circ. 18 of 3 October 1969. See also Report on Marine Science and Technology, 1969 (Cmnd 3992). A more detailed description of the legislative and administrative arrangements is contained in Chapter V of the Third Report of the Royal Commission on Environmental Pollution, 1972 (Cmnd 5054).
46. *The Human Environment—The British View* (HMSO, 1972), p. 14—A Total Approach.

47. Norma W. Humphris, 'Marine Planning', in *Surveyor, Local Government Technology* (19 February 1971), p. 38. See also Royal Commission on Environmental Pollution Third Report, 1972 (Cmnd 5054), pp. 7–10 and Chapter V.
48. A. J. O'Sullivan, *Some Aspects of the Hamilton Trade: Oil Spill in Water Pollution by Oil*, Proceedings of a Seminar held in Aviemore, 4–8 May 1970 (Institute of Petroleum, London, 1971), pp. 315–16.
49. See UN Document E/5003, p. 61, para. 149.
50. Federal Water Pollution Act, as amended by s. 1; Public Law, pp. 87–88.
51. The Canada Water Act of 1970 in conjunction with the Arctic Waters Prevention Act of 1970.
52. See UN Document E/5003, p. 63, para. 155.
53. The Marine Protection Act (Bill) of 1971 (HR 4723).
54. E/5003, op. cit.
55. The United States, Canadian and United Kingdom Acts all incorporate the amendments relating to the abolition of prohibited zones and the stricter definition of 'oil'.
56. Law No. 64–1331 of 26 December 1964.
57. Decree No. 64–412 of 5 May 1964 and Decree of 12 April 1965.
58. Adopted at Brussels on 29 November 1969—IMCO, London Sales No. 1970.3.
59. Water Quality Improvement Act of 1970 as s. 11 of the Federal Water Control Act, sub-s. 6(1).
60. Ibid., s. 11/b(3).
61. Ibid., s. 11/b(4).
62. Ibid., s. 11/c(1).
63. Ibid., s. 11/(d).
64. Ibid., s. 11/(e).
65. Ibid., s. 11/(f).
66. Ibid., s. 11/p(1).
67. Ibid., s. 11/p(4).
68. The Canada Water Act of 1970.
69. Arctic Waters Prevention Act 1970, s. 2(n).
70. Federal Water Control Act (as amended), s. 12.
71. UN Document E/5003, p. 62, para. 157. In Norway the legislation provides for the inspection of conditions in respect of pollution prevention and the responsible Minister is given power to issue instructions to licensees. A licensee is liable in tort for damage or inconvenience caused by his operations and he may be required to post a bond to cover his responsibilities and liabilities. The licence may be revoked for cases of serious and repeated violations of the provisions of the Act or conditions laid down. (Act No. 12 of 21 June 1963 and Royal Decree of 25 August 1967.)
72. Federal Water Pollution Control Act as amended, s. 11(f), sub-ss. 2 and 3.
73. UN Document E/5003, p. 64, para. 159.
74. On this generally see Michael Hardy, 'International Control of Marine Pollution', in a *Collection of Essays in Memory of John McMahon* (Royal Institute of International Affairs, 1971).
75. The United States in notes dated 9 and 15 April 1970 protested against the Canadian proposals. In its reply dated 16 April 1970 the Canadian Government rejected the notes and rejected a proposal that the case be submitted to the International Court of Justice. The Prime Minister of

Canada speaking in the Canadian House of Commons declared that Canada would not go to court until such times as the law catches up with technology (*International Legal Materials* (May 1970), pp. 605–11).

76. Oscar Schachter and Daniel Serwer, 'Marine pollution problems and remedies', *AJIL* (January 1971), p. 84.

77. Ibid., p. 104.

78. Guides and Standards have however been issued by the IAEA on a number of aspects of dealing with radio-active substances. Among these are the Guides and Standards on Radio-active Waste Disposal Into the Sea (Safety Series, No. 5, 1961). The WHO has also issued certain guides in respect of toxic chemicals. See the Annex to the Report of the Secretary-General of the United Nations to ECOSOC 'Problems of the Human Environment' UN Document E/4667 (26 May 1969).

79. Oscar Schachter and Daniel Serwer, op. cit., p. 105.

80. UN Document E/5003, p. 60, note 98. See also *The Times*, 7 July 1971.

81. *International Legal Materials* (March 1972), p. 262.

82. Articles 1 and 4.

83. Article 3.

84. The text of this draft Article is contained in document A/CONF. 48/8/add. 1. The recommendation on the draft is contained in Recommendation 86(d) of the Conference.

85. Text in document A/AC.138/SC.III/L.17 of 24 July 1972.

86. The amendments include also a new and more stringent definition of oily mixture which reads: 'Oily mixture means a mixture with any oil content.' The 1962 definition reads: 'Oily mixture means a mixture with an oil content of 100 parts or more in 1,000,000 parts of mixture.'

87. Oscar Schachter and Daniel Serwer, op. cit., p. 93.

88. See document A/AC.138/SC.III/L.15 of 21 July 1971.

89. *International Legal Materials* (March 1970), p. 25.

90. Article 1.

91. Council of Europe Document 2697 (January 1970), *International Legal Materials*, p. 359. United Kingdom Treaty Series No. 78 (1969), Cmnd 4205.

92. Attachment 3 to the Final Act of the International Legal Conference on Marine Pollution Damage, 1969 (IMCO, London): IMCO. 1970:3.

93. This Group is sponsored by six specialised agencies in the United Nations system namely the FAO, the IAEA, IMCO, UNESCO, WHO and WMO. The Group has conducted studies in various aspects of marine pollution. An indication of the range of the Group's work can be gathered from the Report of its second session GESAMP II/11 (20 June 1970).

94. General Assembly Resolution 2749 (XXV) dated 17 December 1970, operative paragraph 11.

95. Text in document A/AC.138/SC.III/L.17 of 24 July 1972, p. 4.

96. Committee on the Peaceful Uses of the Sea-bed and Ocean Floor beyond the limits of National Jurisdiction—Sub-Committee III.

97. United Nations Treaty Series, Vol. 450, p. 82.

98. UN Conference on the Law of the Sea Resolution II: Pollution of the High Seas by Radio-active Materials, Document A/CONF. 13/L.56. Reproduced in Official Records, Volume II (A/CONF. 13/38 at pp. 143–4).

99. For detailed accounts of the activities of the IAEA in this field, see

Annex XI of UN Document E/4487. Also F. N. Bowder and P. J. Parsons —Control of Radio-active Waste Disposal (IAEA Legal Series No. 5, p. 231). On the reason why IAEA has not been able to take regulatory action, see Paul Szasz: The Law and Practices of the IAEA (1970), IAEA Legislative Series No. 7, s. 23-3, pp. 714–16.

100. UN Document E/5003, p. 58, para. 139.
101. ENEA: Radio-active Disposal Operation into the Atlantic, 1967.
102. United Nations Treaty Series, Vol. 480, p. 43—Article 1.
103. UN Document E/4886 April 1970, para. 36.
104. Text of Draft Treaty in *International Legal Materials*, p. 392.
105. *International Legal Materials*, p. 45.
106. Belgian Ministry of Foreign Affairs and Foreign Trade, Treaty Section, Brussels 1969.
107. International Convention on the Establishment of an International Fund for Compensation for Oil Pollution Damage, 1971: *International Legal Materials* (March 1972), p. 284.

Chapter 10

Pollution by Noise: Social and Technical Aspects

by CHARLES WAKSTEIN
The Open University

Introduction

Noise 'pollution'. Noise is different in a fundamental way from other well-known disruptions of the environment and the phrase 'noise pollution' tends to conceal this difference. In the case of air and water pollution there is an 'external effect' or 'externality' [1]; large-scale damage is done by a polluter to the members of the public without there being any economic relation between them, or any market mechanism that would tend to mitigate that damage. If this view is taken of noise pollution, namely that a pollution is mainly characterised by its external effects, then the extent of the damage caused by noise becomes confined to the annoyance caused by, say, the factory to its neighbours and the annoyance caused by transport noise to people on the ground.

This view, curiously enough, is held by economists who clearly recognise the failings of the market mechanism in the case of air and water pollution but do not recognise the same thing inside a noisy factory, although it has been known since Biblical times [2] and has been recorded by the Greeks [3] that working long enough in a noisy enough environment makes people deafer and deafer. Indeed the number of people made deaf by their jobs is, by any reasonable definition, of the order of 5 per cent of the working population of a developed country. Perhaps the reason for this blindness among economists is that the men and women working in noisy jobs are inconveniently not faithful to the model of 'economic man': informed, rational and free to choose. This author argues that hearing damage may be properly regarded as the most serious damage caused by noise.

On the other hand noise is *similar* to other forms of pollution in that once hearing damage has been caused in people it stands between them and their auditory environment like a cloud between people and the distant hills, worse because the barrier stands between them and their friends. Their friends may be talking directly to them and they can hear that talking is going on but cannot understand the words. This is because the peculiarity of this kind of hearing damage is that the consonants are lost and, with them, comprehension. Moreover, unlike air pollution by smoke, the barrier doesn't go away when the noise goes away; the hearing damage is sometimes irreversible. Like emphysema of the lungs, which limits the use or enjoyment people can get out of their environment, damage to hearing limits the extent to which people can experience their environment.

Value judgements. The technical and economic study of noise presents engineers with new kinds of challenges because it involves them in questions for which there are no purely technical answers. This author found that the most fruitful discussions of these questions were with people from other disciplines. Engineers rarely talk to people working in such fields as politics, psychology, philosophy or theology [4].

For example, when noise-induced hearing damage causes people difficulty in understanding conversation, an important question is how much meaning is lost; this is a philosophical question. Similarly, when one asks whether there are human values that ought not to be quantified that is also a philosophical question.

Further measurements of the effect of noise on the way people perform various tasks fall conventionally in the area of psychological measurements.

Finally, decisions made about 'safe' levels of noise for conservation of hearing are political decisions because, of the people exposed to those levels of noise, a minority will suffer more severe damage than the others; hearing damage is distributed in a statistically normal curve even if the exposure is the same, as may be seen in Figure 10.1. Questions of protection of minorities are political questions.

Moreover, it soon became clear that answers to questions like these reflected implicit value judgements and that even in 'conventional' engineering implicit value judgements are also made. Not only do they occur in deciding how low the level of risk is to be for advanced systems like manned spacecraft, they also occur in designing bridges; if the designer selects a high factor of safety then the risk to people is less and vice versa: thus, the choice of the factor of safety carries with it an implicit value judgement, and this happens in every engineering decision, even in the design of a bolt or nut.

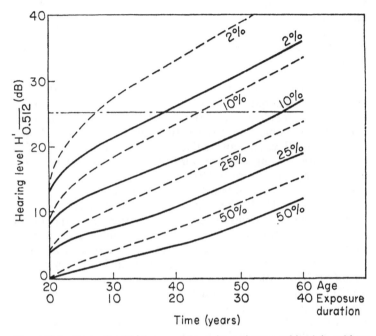

Figure 10.1   Example of 'risk curves' for noise and age combined, in subjects free from other impairments. The ordinate is the average hearing level at 0·5, 1 and 2 kHz. The AAOO 'beginning mild impairment fence' at 25 dB (ISO) is shown. Each pair of curves signifies the level of the occupational noise: 90 dB(A), continuous line; 95 dB(A), interrupted line

Although it may seem a platitude it is the experience of this author that engineers, and economists though perhaps less so [5], find these implicit value judgements an embarrassment and for that reason, it is important to emphasise to them that they can not be avoided.

There are two other political questions: *who* makes the decision about the safe level of noise, and *how* is the decision made? These relate to the important question of participation in decision making [6].

Is noise a serious kind of pollutant?   No, not in terms of the relatively small number of people annoyed by airport noise; but if we take into account the harm that sonic booms might cause in terms of dream deprivation, and the effects of traffic noise in terms of the large number of people living in cities who do not have normal sleep because of this noise, the problem becomes one of greater seriousness.

Occupational noise without any doubt *is*; if we consider things that make people blind as serious then industrial deafness is serious.

Why isn't more done about it?   Part of the reason is that people are uninformed about the risk, even those working in noisy jobs [7], according to people active in the field in the US;* and the more one talks to people, the stronger this impression gets. A British mineworker [8] was not aware that the noise levels at the coal face could cause hearing damage, and union officials [9] were less informed than management [10].

People are worried about the wrong sources. Rock music can cause deafness [11, 12] but the exposure is rarely long enough. Very intense sonic booms can cause hearing damage [13] but these are extremely unlikely to occur. The same applies even more to subsonic jet noise outside houses very near airports; although the noise may seem terribly loud, the maximum level is only about 90 db(A). This could cause hearing damage in a minority of people but only if they were outside continuously. There are many misconceptions about this.

It would be helpful if more were known about the exposure of people to dangerous noise levels. Part of the argument in an important recent British case, the Berry case [14], revolved around how much of a worker's hearing damage was caused by sources outside his work. It was argued that the employer could not be expected to compensate the worker for damage he had caused himself in his private life.

From the standpoint of the researcher trying to find out how many people are exposed to how much noise for how long and from what sources it would be convenient, so to speak, to follow people around day after day. This has in effect been done by one student housewife in the United States [15] who carried recording equipment around with her all day. However, it is an inconvenience and some people regard it as an invasion of privacy.

It might be thought that having to pay Workmen's Compensation awards, as in the US, would deter industry from allowing

* At present it is not justifiable to say that this is any more than an impression because no studies of awareness have been done.

hearing damage to take place in their workers. However, these awards are paid by insurance policies in the States and presumably would be in Britain as well. As the premiums for *all* workmen's compensation, i.e. other injuries and diseases too, amount to only 1 per cent of payrolls in the States [16], this will hardly be seen as a deterrent. Even if the premiums were to double on account of the insurance company having to pay out a large number of claims, this would not cut into profits seriously. Small as is the threat to industry's profits, things are made even easier for industry by the 'waiting period'.

Workers must work away from noise for a certain time—often six months—before they can legitimately claim that their hearing damage arose from their job [17]. This is reasonable, as it is known [18] that there is some recovery in the period after stopping work in noise. What still appears to be in question, however, is whether the waiting period is much longer than necessary or not. Rosenblith [19] reported measurements of the hearing of men and women exposed to noise for about two years which were made about two days and again about one week after exposure had stopped. The improvement after one week was noticeably greater, as much as 25 dB at 4 kHz. Rosenblith also reported measurements made on a group of people who had on average been exposed to the same noise for a longer time than the first group, more than ten years. In this case, measurements were made two days and again six weeks after exposure had stopped. The improvement was not so marked, the maximum improvement of the group average being 11 dB at 6 kHz; at all other frequencies measured the improvement was on average only about 3 dB and did not exceed 6 dB. This evidence suggests that no measurements were reported for periods longer than six weeks. For long exposure, where the hearing loss is greater, the recovery is less and a waiting period of less than six months may be justified, but research is needed to resolve the question. Or alternatively, with the results of such research, available measurements of hearing may be corrected for the improvement and thus, the waiting period may be reduced to a week or even less.

It would also be helpful if more were known about the cost of industrial noise control. These costs are not easy to obtain from companies with actual experience and although extensive descriptions of actual methods are published in places like the *Handbook of Noise Control* [20], no costs are given. This lack of easily available information together with the fact that cases of really expensive noise control somehow seem to be well known, makes it difficult [21] to convince people that noise control on an industry-wide basis can indeed be very cheap. However, there are encouraging signs in that a few manufacturers in Britain [22] and the States [23] have been willing to make their costs public.

Although as recently as 1970, there were very few articles on hearing damage in the press, encouraging signs have come from the media recently in Britain. The weekly press have shown interest in publishing articles about noise [24]; the BBC [25] have given serious treatment to hearing damage; and the daily press have given coverage to the risk of hearing damage, perhaps because of the Berry case [26].

How might resources be allocated in fighting the various pollutants? Perhaps noise is not a serious problem and other problems like air and water pollution, education and housing need money spent on them more urgently. Having said that, it is easy to fall into the trap of trying to get an idea of how serious the noise problem is by expressing it in monetary terms and then making comparisons with other problems, also expressed in monetary terms. For example, one might take Stone's estimate [27] of the number of men (1 million) in Britain having occupational hearing damage, and apply to it a proposed figure (£7·50 per week for compensation in the case of complete loss of hearing), and then take into account that the average hearing loss [28] will be about 10 per cent. This way, once committed to finding a figure, one finds a figure—in this case £40 million per year. This is something like half the yearly expenditure on Social Security payments to victims of industrial accidents [29]. It is thus not insignificant. These men have taken, on average, thirty years to acquire their hearing damage, so something like 30,000 men a year have been getting deafened. What would be the reaction if people were told that 30,000 men a year were getting blinded* by their jobs? Then monetary comparisons would seem less relevant. In making monetary comparisons, otherwise called quantification of values, it is easy to lose the full appreciation of what the money represents and for that reason and others, quantification can be considered at best a crude tool for decision making and at worst, completely inappropriate.

* People working with deaf people usually consider deafness *worse* than blindness because it cuts people off from each other as a result of their not being able to talk to each other.

Scope, goal and framework of the chapter. This chapter attempts to gather together the latest information and controversy, all of which bear on the technical aspects of noise as a disruption of the environment. These include such disparate topics as technical solutions, quantification of annoyance, exposure to noise, hearing damage, costs of noise control, quality of the environment, and ways of making decisions about noise control.

The goal of the chapter is to study the technical, economic and other difficulties in characterising, measuring and controlling the disruption of the environment caused by noise.

The straightforward technology of noise control is well documented in the literature (see the References, p. 261) and is only briefly referred to here.

The chapter uses for framework a paper based on a study [30] of the noise problem in the US carried out by a group* at Carnegie–Mellon University in the summer of 1967.

The task of the study was to find out how serious the noise problem is in the United States. This was phrased in terms of a number of questions as shown in Table 1. The questions that are underlined are the ones that are not dealt with in the standard references, like the Wilson Report [31], the World Health Organisation monograph by Bell [32], the *Industrial Noise Manual* [33], the *Handbook of Noise Control* [34], the London Airport Study [35], and the Central London Study [36] and the books by Burns [37], Duerden [38] and Kryter [39].

These references, standard until the mid-sixties, have been to some extent superseded by a new crop of references which have adopted a broader orientation. Among this new crop are the CEQ survey [40] and the books by Goodfriend [41] and Taylor [42].

Finally, one of the aims of this chapter is to demonstrate by reference to the work of the C–M group on the noise problem in the United States, that it is possible to arrive at estimates, even rough ones, of the extent of the problem and the scale of the remedies for any industrialised country by using existing data and drawing new conclusions from them. The argument is much the same from country to country, so in this sense, it may be said that data from Britain and the States are really more general.

Exposure to Noise

Relative importance of the various sources. In spite of the great interest in the problem of annoyance caused by jet aircraft noise

---

* Referred to as the C–M group.

near airports, it seems inescapable that hearing damage caused by exposure to noise is more widespread.* Most of this hearing damage is caused by noisy jobs.

It is difficult to decide which aspect of exposure to noise is next most important. The major sources are listed below:

1. *Traffic noise* exposes about 60 million people [44] in US cities to noise levels exceeding those recommended by the ISO [45]. Indeed, Richards has pointed out that the noise from diesel vehicles has increased by 12 dB [46] since 1966, when the Wilson Committee recommended anti-noise legislation, so things are getting worse; at least, they are in Britain. The work by Richards is the main indication that traffic noise may soon be causing a public health problem.

2. *Jet aircraft noise* near airports exposes perhaps 8 million people in the States to noise levels that produce such effects as waking people up, making it difficult for them to get to sleep, disturbing their rest and relaxation etc. It has been suggested [47] that the number of people annoyed by airports could be reduced by wider use of short take-off and landing (STOL) aircraft and that the landing strips for such aircraft could be sited much closer to cities, or even in cities. Recent discussion [48] of the severe annoyance that would be caused by STOL strips in cities have made clear the difficult compromise between economic benefit and environmental disadvantage involved in the siting of such airport landing strips; the cheapest land to acquire in cities is land in or near slum areas. Thus, to entertain a suggestion that many poor people should be subjected to annoying, perhaps severely annoying, levels of noise rather than fewer rich people, or even industrial properties, becomes an expression of political priorities. This is an example of the 'non-technical' considerations or value judgements that are inevitably associated with noise problems.

As for wider use of helicopters, there has been abundant experience of the annoyance this can cause in New York where helicopters land on the roof of the Pan American building on 42nd Street. The people nearby dislike it so much that they have forced the curtailment of night and early morning flights [49]. The

---

* With only one possible qualification, that if such studies as that of Richter [43] show an important deficiency in the kind of sleep people get sleeping in rooms exposed to traffic noise then the number of people failing to get healthy sleep is so large, that even if the individual effect is small, the scale of the harm done is such, that the problem seen as a public health problem is also a serious one.

picture is not good for ways of avoiding the large areas exposed to high noise levels around airports.

3. *Sonic booms* might become a very important source of noise in the future, if there were regular overland flights of SSTs, supersonic transport aircraft. Kryter [50] has suggested that perhaps 50 million people in the States would be exposed to perhaps fifteen booms a day, some of which might be at night.

With cancellation of the American SST, the problem of sonic booms becomes one more for Britain and France as the first two prototype Concorde SSTs are flying, and indeed some sonic boom tests [51] have been done up and down the west coast of England and Scotland. The results of the social survey [52] following these flights have not yet been published. If a few hundred of these aeroplanes are sold (the chances of this do not appear good; all seventy-two options have been allowed to *lapse* [53]) and are used for transatlantic flights over Britain, then the expected area that would be painted with boom carpets is as shown in Figure 10.2 [54].

As far as night booms are concerned, Lukas *et al.* [55] have recently published a report which suggests that although children are relatively insensitive to sonic booms, a very small minority, about 2 per cent, are awakened. Middle-aged men appear to divide up into two groups of high and low sensitivity to noise. Among those of low sensitivity, about 7 per cent are fully awakened if they are in the important sleep stage when rapid eye movements (REM) and dreams occur. Of the sensitives, nearly 40 per cent are awakened during the REM sleep stage. Among older men, who also divide into sensitives and insensitives (no women were tested), nearly 12 per cent of the insensitives and over 50 per cent of the sensitives were fully awakened. As the disturbing results of being awakened regularly during the REM sleepstage are well known [56] there is reason for looking very carefully at any decision to create night-time booms.

Industrial noise. The number of men exposed to dangerous levels of noise, i.e. high enough to cause hearing damage, was estimated by taking the US census list of occupations and the data of Karplus and Bonvallet [57] for noisy areas and noisy machines and matching them up. It was estimated whether the exposure to these noise levels was for a large part of the day, say five hours, or not. This gave 11·6 million workers exposed to 80 dB or more, based on the 1960 census. The 'measured' distribution is shown as M in Figure 10.3.

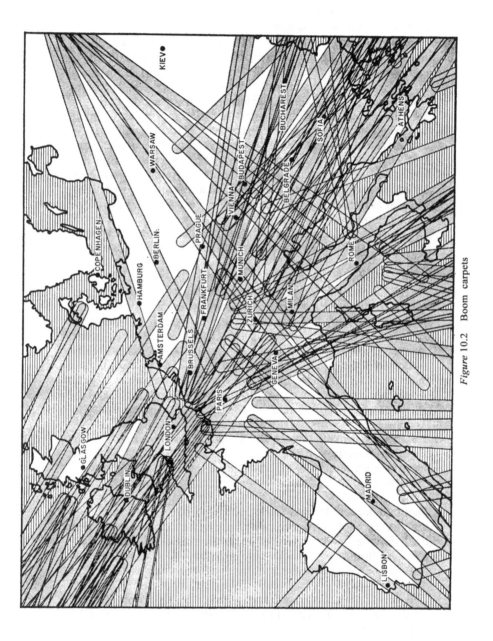

Figure 10.2   Boom carpets

In addition to these jobs, there are a large number of jobs for which the noise levels have not been formally measured, but for which it was possible to tell what the noise sources would be. For these noise sources, one only had to estimate whether the noise levels were likely to be above 80 dB based on comparisons with other noise sources and noisy areas, and to estimate how often the workers would be exposed to these sources. In this way, it was estimated that an additional 22·4 million workers would be regularly exposed to noise levels about 80 dB. This still left the problem of the unknown distribution of noise levels in these jobs. It was estimated in two ways. The first way was based on the assumption that the distribution was the same as for the measured occupations and was rejected as an unjustifiable simplification. The second way gave a lower estimate which was preferred because the jobs for which noise levels have not been measured yet are those that are obviously less likely to have high noise levels. In the second a triangular distribution biased towards low noise levels was used. This gave the measured plus suspect distribution shown as M + S in Figure 10.3. The important conclusion to be drawn from the figure seems to be that there are about 8 million workers exposed to noise levels above, say, 95 dB in their work.*

The figure of 8 million is greater than an estimate of the number of people with hearing damage as defined by the American Academy of Ophthalmology and Otolaryngology (AAOO) criterion [58]. This was estimated to be 4·9 million [59] by extrapolating the Wisconsin State Fair data [60] to the US.† This means either that the 4·9 million figure is too low or the 8 million figure is too high, or to be more pessimistic that both figures are

* The C–M group chose 95 dB as a typical figure because the ISO damage risk criterion [61] recommends that workers exposed to 87 dB(A) for more than five hours a day be protected and because the A-weighted noise levels are not uncommonly 8 dB lower than the unweighted C-scale values.

† A similar extrapolation has been made for Great Britain [62] (the figure was 1,204,000) and was criticised on the grounds that the people who volunteered to have their hearing measured at the Wisconsin State Fair were not a random sample but a self-selected sample of the men and women of Wisconsin. However, even if this resulted in an overestimate of the number of people having industrial hearing damage, the number would not be overestimated by a factor of ten. Thus, even taking a figure of a tenth as many the number of people in Britain at work is probably at least 120,000. This figure of 120,000 has been criticised as too low and a figure of 1,000,000 suggested by Acton [63]. More recently, Acton's figure has been confirmed by a government spokesman [64] and even this figure is an underestimate [65] as the Factory Inspectorate does not cover jobs in mining and agriculture which are known to be high risk jobs.

too low, or for that matter, that they have not worked long enough at these jobs for them all to have suffered hearing damage.

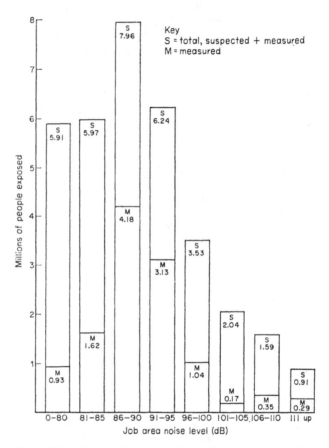

Figure 10.3   Estimated exposure of American workers to noise

There is an inherent problem in making such estimates that if at a given time one finds a certain number of workers in a certain job their hearing damage may not have arisen only from that job. Clearly, it is desirable to be able to check this figure. As it happens, although there are about ninety occupations for which one suspects noise levels may be high, the great majority of the 22·4 million workers, in fact 18·6 million, are found in only twenty-five of these occupations. This suggests that noise surveys be made in certain occupations. If two weeks were needed per occupation, it would only take a year to complete the measurements with one set of instruments. Thus, one could get a lot of information without large-scale effort.

Information on noise exposure in other countries is limited: of the delegates to the 1970 NPL Conference from ten different countries, only two were able to provide an estimate of exposure, Christensen [66] reported that for Denmark the fraction of industrial workers exposed to levels above 90 dB(A) was 3 per cent and Mme Passchier-Vermeer [67] from the Netherlands estimated that the fraction exposed to levels above 80 dB(A) in the Netherlands was 3–5 per cent. Respondents from Spain, Switzerland and Sweden could not give any estimates. It is interesting to note that whilst 90 dB(A) is considered a safe level by the British [68], people in the Netherlands [69] prefer a safer level of 80 dB(A).

Finally, there are obviously other sources of noise and the Wilson Report [70] has documented these very well. The main ones, however, are the three dealt with below.

Traffic noise. To evaluate the noise from traffic to which people are constantly exposed, Table 1 was developed for the United

Table 1. *Estimated exposure to traffic noise in urban residences in the US*

Vehicles per day (thousands)	Type of traffic	% of population exposed	Number of people exposed (millions)	Noise level inside houses due to freely flowing traffic (dBA)		
				11 p.m. to midnight	Average for the day	During worst rush hour
5	Unspecified	73·5	92·8	29	31	38
5–10	Lorries and cars	2·0	2·5	54	56	65
5–10	Mainly cars	4·0	5·0	42	44	51
10–20	Lorries and cars	8·5	6·6	49	51	58
10–20	Mainly cars	5·5	6·9	46	48	55
20–40	Lorries and cars	5·0	6·3	57	59	66
20–40	Mainly cars	0·5	0·6	54	56	63
40	Lorries and cars	0·5	0·6	59	61	68
40	Mainly cars	0·5	0·6	56 (ISO) (25)	58	65 (ISO) (35)

States from traffic data of the Boston Redevelopment Authority [71]. The table shows the percentage of population exposed to various traffic volumes. These traffic volumes and an empirical relation presented by Bolt, Beranek, and Newman [72] were used to calculate the noise levels that would be produced during the worst rush hour, and the levels between 11 p.m. and midnight when people might want to go to sleep. According to the ISO*

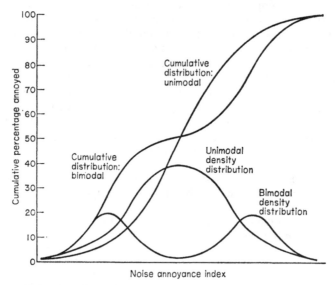

Figure 10.4   The effect of the shape of the annoyance distribution on the choice of an acceptable level

* The ISO standard, in common with other human factors or ergonomic standards, is based on a unimodal probability density distribution or response. Such is shown in Figure 10.4. The cumulative probability distribution, the integral of the density distribution, is also shown. If on the other hand the density distribution is bimodal, and there is abundant evidence that this is the case both for noise annoyance and for noise induced hearing damage, then the density distribution and the cumulative distribution will be as shown in Figure 10.4. The important implication of this is that if a level standard is set for the annoyance of a minority on the assumption that the density distribution is unimodal and it turns out that the density distribution is bimodal then the annoyance will be seriously underestimated; a level thought to annoy only 10 per cent may well annoy 40 per cent. Moreover, the corroborating evidence based on experience is usually based on the level of complaint and this neglects those who were very much annoyed but did not complain. McKennell [76] makes this point very well, in the London Airport Study. There are strong hints of this bimodality in the work of Bryan [77] on annoyance.

standard of annoyance [73], daytime noise inside a dwelling in an urban residential area should be no more than 35 dB(A) if the windows are closed. When one compares the volumes in the table with the levels recommended by the ISO, one sees that the noise inside homes can be at a level which is constantly annoying. From these data were extrapolated the number of people exposed to similar traffic volumes in cities throughout the United States. The traffic noise measurements were made in specific parts of Boston that accounted for about 40 per cent of the total 1960 population of the city and approximately 25 per cent of the area. If one assumes that Boston is like other American cities in population density, car ownership, and traffic flows and if one applies the upper 40 per cent figure to the 150 million urban population of the United States, one gets a figure of 60 million people annoyed by traffic noise. If one applies a more conservative figure of 25 per cent to take into account the fact that the newer cities of the United States like Los Angeles are more spread out, one gets a smaller figure of about 38 million people annoyed.

Traffic flows may be distributed differently in Boston but specific comparisons and extrapolations are difficult to make and in the absence of such extrapolations, the above estimate is all that can be made.

One should take into account that people in the remaining 60 per cent or 70 per cent of the urban population are also exposed to less severe traffic noise in their homes, but it is not known how much less severe.

Car ownership in Boston is about average for cities in the United States [74] so that this aspect can be extrapolated safely.

The calculations are based on average freely flowing traffic volumes and have not been refined to take into account stop-go traffic, or traffic on hills, which would be louder [75].

These calculations can be compared with similar ones made in planning a motorway (M62) through urban areas in Liverpool. The noise level assumed to be acceptable inside a house in the daytime was 45 dB(A) based on the Wilson Report. This is 10 dB(A) higher than ISO. So you pays your expert and takes your choice and to hell with subtleties such as unimodality or bimodality. Thus, there is a value judgement even at this early stage in the chapter. Although one may disagree with these figures, one agrees with the following observations made in the MALTS (Merseyside Area Land Use Transportation Study) Report [78].

1. The data relating noise levels produced on different types of roads and by different volumes of traffic is limited and crude.

2. Knowledge of the social nuisance due to traffic noise and of the acceptable levels of noise is in an early stage of development . . .

Jet aircraft noise. In estimating the number of people exposed to jet aircraft noise near airports, the London Study [79] is especially useful as it is one of the very few published on commercial jet aircraft.* In the London Airport Study essentially all of the people annoyed by the jet aircraft noise were found in a ten-mile radius of the airport. Annoyance was defined by an average annoyance score of 3·5.

'The score of 3·5 corresponds to the average reaction at the highest noise exposure level (103 + PNdB), falls mid-way between a self-rating of "moderately" and "very much" annoyed, is the point beyond which—on various indices—aircraft becomes singled out as the greatest inconvenience of living in the area, and at which 50 per cent of informants report disturbances to sleep and rest' [80].

It is possible to obtain a preliminary estimate [81] taking the two dozen major air traffic hubs in the United States [82] and counting only those for which the airport, like London Airport, was ten miles or less from the centre of the city. One then assumes that half the population of the city lived in the area around the airport. Using the ten-mile figure meant that one discounted Kennedy Airport, O'Hare Airport and Los Angeles Airport, so it is clearly an underestimate of the number of people exposed. The total is nevertheless 8 million. We discuss later on how many of these 8 million people are annoyed.

* For comparison the more recent French study [83], the British study by Mil Research [84] and the American Tracor study [85] and the Dutch study [86] indicated that people are annoyed (annoyance being defined as reporting interference with face to face conversation) as follows: London 41 per cent, American airports 37 per cent (Denver) to 66 per cent (Los Angeles). The definition of annoyance in the 1961 London Airport study by McKennell was based on 50 per cent of those interviewed reporting disturbance to sleep and rest. By this criterion, Americans appear to be less annoyed by their airports: the percentage of those reporting disturbance to sleep being 57 per cent for London and between 19 per cent (Denver) and 35 per cent (Los Angeles). This points up the difficulty of comparing the annoyance caused by different airports. Nevertheless, compared with other airports in the States and with London, Los Angeles must be one of the worst airports in the world to live near. It scores worse on every type of disturbance than any other American airport, and better than London only on general aircraft annoyance and the percentage of people *not* annoyed.

In the short space of a few years since these estimates were made, three major changes in the way people in the field look at these problems have occurred:

(a) PNdB* has been modified to take into account the duration of the flyover [87] and the effect of pure tones in the noise [88]: this applies especially to the landing noise when the characteristic whine caused by the compressor blades is heard.

(b) The validity of NNI* as a measure of annoyance has been questioned in a British study [89]; it has been suggested that PNdB is the more relevant index and that the number of flights in a day (or night) is irrelevant. This conclusion is challenged by Alexandre [90].

(c) Doubts have been expressed [91] about saying that there is a Noise and Number Index below which no one is annoyed. These doubts have been given stimulus by the appearance of the Roskill report [92] on the siting of the Third London Airport which took as the noise affected area only that within the 35 NNI contour. The 35 NNI contours just avoid towns but lower contours include whole towns. There is clearly a value judgement involved here too. And the value judgement is an important one because the cost of annoyance would be greatly increased if the threshold NNI value were lowered.

Aside from that, however, the important Tracor study [93] shows for the first time that NNI, or for that matter, any purely physical index, is inadequate to predict annoyance; certain personal factors, independent of the activities interfered with, explain the majority of the variations in annoyance.

Sonic booms. For the United States one may accept Kryter's estimate [94] of 50 million people exposed to fifteen booms a day. One rather suspects it is high because one recognises 50 million as the rural population of the United States and because the SST overland flights would be mostly over rural areas and the sonic boom carpets, as discussed later, will not cover a majority of the area of the United States. Thus, if the figure is correct, one implication one can draw is that cities will be subjected to sonic booms. This suggests that deviations from the planned routes, even small ones, may make a big difference, since according to

* PNdB = Perceived noise in decibels; NNI = Noise number index.

Maxwell one may assume SST routes will not include detours around cities [95]. As the present system of navigating overland flights by radio beacons either manually or by automatic pilot results only rarely in excursions outside the ten-mile-wide air traffic corridor [96] the most conservative assumption to make is that the accuracy 3σ (r.m.s. error = 1σ) is ± 5 miles. This will clearly have to be improved considerably to avoid subjecting cities, which are for the most part smaller than that, to sonic booms and the improvement will have to go beyond what can be achieved at present by inertial navigation systems. Brown [97] reports that the accuracy for such systems is (2σ = ± 3 miles). Concorde has a dual inertial system of higher accuracy.

How Serious are the Effects?

Hearing damage. One might think that it would be straight-forward to work out how serious it is for there to be 4·9 million people in US society with hearing damage caused by noisy jobs. However, it was discovered very early in the C–M study [98] that a number of value judgements are implicit in any such assessment.

It was noticed that at the NPL symposium* in 1961 [99] there was a certain amount of disagreement about whether being able

* It is interesting to compare the attitudes implicit in the discussion that took place in the 1961 NPL conference with those that took place in the 1970 NPL conference. It might have been expected that with the appearance in Britain of groups like British Society for Social Responsibility in Science, more discussion would have been centred on questions of human values.

To a certain extent this has happened; in quite practical terms the standard for measuring hearing damage now recommended by the British Occupational Hygiene Society [100] is even more liberal than the Australian method of Murray [101] as it takes into account measurements at six frequencies including 3 kHz, 4 kHz and 6 kHz at which the hearing damage is more pronounced.

But a polarisation has also taken place; together with such discussion were incidents like the question asked by a company medical officer [102] whether his responsibility was to tell workers coming to see him that if they continued to work in areas of high noise levels they would lose their hearing or whether his responsibility was to his employer and thus was to avoid telling such things to workers. It is easy to say that this is an erosion in personal responsibility but it must be understood that the present structure of the industrialised society in which this doctor finds himself is such as to bias such decisions; the question is perhaps more that the society itself should be such that people, especially doctors, will not be placed in a position where these pressures are exerted. Similarly a conference organiser rejected a suggestion that reporters from a weekly journal be invited so that the problem of industrial hearing damage could be publicised.

More generally, things are not helped by the 'two cultures' attitudes of such journals. The journal in question thought that an article on hearing damage was too 'scientific'. The editor of the proceedings also left out the important to hear, say, music without severe distortion or being able to understand speech should be the overriding consideration in hearing conservation. So a filter was set up to simulate an early stage of noise-induced hearing loss for which Workmen's Compensation would not be awarded. Much to the surprise of the group the difference was quite obvious. Now this is misleading for two reasons. First, since noise-induced hearing loss progresses slowly, the person has a chance to become accustomed to the way things sound. And second, there is an exaggerating effect such that things actually sound worse to the person with the hearing loss because of the phenomenon of masking by adjacent tones [103]. Even hearing aids are not satisfactory [104]; selective amplifications of the affected frequencies will still not make speech comprehensible because in addition to masking there is also recruitment* [105] and unwanted amplification of the background noise. Three of the participants in the 1970 NPL conference [106, 107, 108] confirmed that these difficulties are not being dealt with successfully in the present-day hearing aids. One may say that with noise-induced hearing damage the quality of the environment, as perceived by the person, has been degraded. Thus, if this degraded quality of the environment is unacceptable, then there has to be some way of characterising it so that it is possible to tell that this unacceptable condition has been reached. In other words, an assessment scheme has to be found that will say that there is an impairment *before* there is any noticeable loss in speech comprehension. The well-known AAOO method [109] uses only the frequencies 500, 1,000, and 2,000 Hz. From the experience of the C–M group with filtered sounds this is inadequate if quality of the environment is to be taken into consideration. The only assessment scheme found in the C–M study which uses 4,000 Hz was the Australian one described by Murray [110].

More recently, however, at the 1970 NPL conference Atherley and Noble [111] described assessment schemes like the AAOO scheme, based on taking the average loss at three frequencies as well as schemes based on taking the average loss at 4, 5 or even 6

---

discussion that made industry's attitudes clear (they did not believe that control could be cheap) even after written contributions had been submitted.

A suggestion that, as it was a 'star-studded' conference and the question of awareness had arisen it might have been useful to have made a press statement, provoked no response, either for or against.

* At a given frequency the margin between audibility and discomfort is very much reduced, so the person with noise induced hearing damage will say, 'Louder, I can't hear you', and if the speaker does talk louder, will say, 'Not so loud, don't shout'.

frequencies. The assessment of the damage based on such schemes was correlated with the assessment based on psychological interviews evaluating people's perceptions of their handicaps. These were classified in terms of speech hearing, non-speech hearing, localisation, acuity for non-speech sound, emotional response to hearing loss, speech-hearing distortion, and personal opinion of hearing. It was found that the assessment based only on pure tone audiometry did not correlate with any of the categories, no matter how many frequencies were used in the averaging.

Atherley and Noble reached the important conclusion that 'an adequate clinical picture cannot be obtained from pure-tone thresholds alone'.

A different assessment scheme used in Austria which uses speech audiometry rather than pure tone audiometry was described at the 1970 NPL conference by Raber [112]. A similar method based on speech audiometry is also used in Switzerland [113].

The number of people identified as having hearing damage depends very strongly on the assessment scheme one uses. If one uses the AAOO scheme, one concludes that there are 4·9 million people in the States having hearing damage. However, if one uses the Australian scheme, one concludes that there are 14 million people. Thus the choice of a scheme for assessing hearing damage is, in effect, a value judgement.

One way of expressing the seriousness of hearing damage in such a large number of people is to assign a monetary value to the percentage of impairment. For the States this may be done in terms of Workmen's Compensation awards.* The minute one looks at the variation in Workmen's Compensation awards from state to state, as shown in Table 2, one has to ask which one is 'right', if any. Again, the one that is chosen reflects a value judgement. Assuming the third quartile figure to be representative, the dollar amounts become $6 billion by the AAOO scheme and $12 billion by the Australian scheme. The average length of exposure for the AAOO scheme is thirty-five years [114], so that the claims are building up at the rate of $150 million a year, which is £67 million a year. By the Australian scheme the average length

---

* There is, of course, the question of whether the awards represent the economic and psychic loss to the society. Expressing psychic loss in monetary terms poses a vexing and important question, quantification of values, which has been treated by Lave and Seskin [116], Weisbrod [117], Dawson [118], Blum [119], Heymann [120], Marlowe [121] and the present author [122] among others.

The particular monetary figure chosen reflects an implicit value judgement as indeed does the very choice to quantify the loss in monetary terms.

of exposure is, say, thirty years [115], and the claims are building up at the rate of $400 million a year, which is £170 million a year.

Awareness of risk of hearing damage. The C–M group were surprised that the estimates of the number of people in the States exposed to risk of hearing damage came out so high. They wondered whether workers were aware of the danger and when they found tree workers blithely exposing themselves to 120 dB(A) from their chain saws, they started making a special point of asking people in the field about awareness when they talked to them. The opinions were essentially as follows:

Assistant Professor, Department of Speech and Drama: Old-timers tell newcomers about the danger of hearing damage in boiler works, forge shops and foundries.

Union Official, Director of Health and Safety: Men working in noisy industry do not know of the danger of hearing damage.

Director of Industrial Hygiene, Primary Metals Producer: Awareness is limited to the large enlightened companies that have hearing conservation programmes.

Two Officials in State Division of Occupational Health: Some unions in notoriously noisy industries do know of the problem, but in general awareness is limited to the large enlightened companies. For instance, in the State the total number of manufacturing companies is 18,000.

Only one-sixth of the plants, that is, the ones with more than 100 employees, for which the average number of employees is 250, have the capability of giving this information to their workers because they have well-established industrial hygiene programmes. It might be said, therefore, that five-sixths of the manufacturing companies in the State, as an example, have a low level of awareness of the danger to hearing caused by exposure to noise. This, however, is somewhat mitigated by the assistance given by the insurance companies that carry Workmen's Compensation insurance for these companies.

State Principal Industrial Hygienist: His State has an active programme which involves visits to plants, measurements of noise levels and recommendations to management. Where these visits are made, the level of awareness is high. On the other hand, this official has not encountered one case in his long

Table 2. *Comparative provisions for noise-induced earing loss in american workmen's compensation cases [251]*

State	Is noise-induced hearing loss compensable?	Basis of compensation schedule in weeks		Maximum dollar amount payable		Must employee leave work to file claim?	Test frequencies and rating scales to determine hearing loss and impairment
		one ear	both ears	one ear	both ears		
Ala.	No	50	150	1,650	4,950		ME
Alaska	Yes	52	200	1,800	7,000		AAOO & AMA
Ariz.	No	85	255	11,000	33,000	No	AAOO & AMA
Ark.	Yes	40	150	1,400	5,250		ME
Calif.	Yes	*	*	1,408·75	5,594·75	No	*
Colo.	No	35	139	2,860	8,580	No	AAOO & AMA
Conn.	Yes	52	156	3,750	8,750	No	ME
Del.	Yes	75	175	3,640	14,000	No	ME
DC	Yes	52	200	1,680	6,300	No	AAOO & AMA
Fla.	Yes	40	150	2,220	5,550	No	ME
Ga.	Yes*	60	150	5,850	22,500	No	ME
Hawaii	Yes	52	200	1,050	4,500	No	AAOO & AMA
Idaho	Yes	35	150	*	*	No	AAOO & AMA–ME
Ill.	Yes	50	125	3,150	8,400	No	ME
Ind.	No	75	200	1,850	6,500	No	ME
Iowa	Yes	50	175	1,140	4,180	No	ME
Kan.	Yes	30	110	2,080		No	AMA Formula Feb. 1947
Ky.	No			Depends on wages			ME
La.	No			1,950	3,900	Yes	ME
Me.	No	50	100	1,875	4,375	No	ME
Md.	No	75	175	3,000	8,000		ME
Mass.	No	150	400				ME
Mich.	Yes					Yes	ME
Minn.	Yes	55	170	2,475	7,650	No	ME
Miss.	Yes	40	150	1,400	5,250	No	ME

State							
Mo.	Yes	17	100	765	4,500	Yes (6 mo.)	AAOO & AMA–ME
Mont.	No	40	200	2,000	10,000		AAOO & AMA
Neb.	No	50	100	1,850	3,700	No	ME
Nev.	Yes	85	255	2,000	6,000		ME
NH	No	52	114	2,184	4,788		AMA Formula Feb. 1947
NJ	Yes	60	200	2,400	8,000	No	ME
NM	Yes	40	150	1,520	5,700	No	AAOO & AMA
NY	Yes	60	150	3,300	8,250	Yes (6 mo.)	ME
NC	No	70	150	2,450	5,250	No	AAOO & AMA
ND	Yes	50	200	1,575	6,300	No	ME
Ohio	Yes	*	*	1,225	6,125	No	ME
Okla.	No	100	200	3,000	6,000		ME
Ore.	Yes	60	192	2,790	8,928	No	ME
Pa.	Yes		180*				ME
RI	Yes	60	150	1,620	4,050		ME
SC	No	70	150	2,450	5,250		AAOO & AMA
SD	No						ME
Tenn.	No	75	150	2,700	5,400		AAOO & AMA
Tex.	No		150	?	?	No	AAOO & AMA
Utah	Yes	50	100	2,100	4,200		ME
Vt.	No		215		7,740	No	ME
Va.	Yes	50	100	1,850	3,700	No	AAOO & AMA
Wash.	Yes			1,950	6,825	No	ME
W. Va.	Yes	60	180	2,280	6,840	No	AAOO & AMA
Wis.	Yes	36	216	1,539	9,234	Yes (6 mo.)	ME
Wyo.	No	*	*	2,500	5,000	Yes	
Can. Prov.	Yes			11-25	112-50	Yes	ME
				(per month for life)			

The problems associated with noise induced hearing loss are not static, and the rules and provisions pertaining to same are subject to change as new knowledge is acquired and as agreement is reached regarding legal interpretation and compensation matters. Furthermore, the ruling schedules and benefits paid for loss of hearing are subject to change at any time by administrative or legislative actions. It is therefore suggested that the reader consult the existing compensation acts and schedules of the states or provinces involved in order to detect any change in the provisions of the above survey made in 1963 [249].

ME indicates medical evaluation.

AAOO indicates Reference [58], AMA, Reference [248].

* Special case.

career where noise control was a major issue in collective bargaining.

Noise Consultant: Awareness is low.

Professor and Chairman, Department of Economics: There must be a tradition in noisy industries that the oldtimers would tell newcomers.

This opinion was shared by some of the members of the C–M group. The question of awareness in Britain has been discussed by Bryan [123] in a paper presented at the 1970 NPL conference. In the discussion following the author urged that the important question of communication with the public should not be set aside. This question was also discussed at length in the Design Participation Conference in Manchester in 1971 [124].

At the 1970 NPL conference, Chadwick [125] also pointed out that people in Britain were very little aware of the risk of hearing damage from shooting guns. A close friend of the author developed hearing damage from target shooting with a pistol over a period of five years in spite of the fact that he wore earmuffs which were claimed to be adequate. He gave revealing anecdotes of the difficulties caused by such hearing damage; in his job interviewing hospital patients he had difficulty distinguishing between 'fifteen' and 'sixteen' to such an extent that tempers became frayed. He could not hear his wife whispering intimately to him in a cinema. He was unable to hear a car driving into a gravel driveway twenty feet away from him on a day when the wind was blowing gently in the tree in his back garden.

This was borne out in a more general way in the famous (in Britain) Devon Assizes case described by Coles [126] where a man who suffered hearing damage from firing an explosive stud-driver failed to win his case because it was ruled that knowledge of the risk was not general enough. It was suggested by Wakstein [127] that the risk of hearing damage from impulsive noise such as drop-forging noise had been known of since the 1940s. Coles [128] in reply was careful to point out that the pressure histories of drop-forging noise and proof-firing of guns are different from the pressure history produced by an explosive stud-driver, so that the risk in one situation does not necessarily inevitably mean risk in another.

Such careful distinctions are what one traditionally expects from a professional, but perhaps it would be more useful to say instead: 'The two noises aren't the same, so we can't definitely say the risk is the same *but* the noises are both impulsive, and impulsive noises are notoriously dangerous (even children's toy guns are dangerous) [129]. We ought to *assume* the stud-driver is dangerous *until* we have carried out enough measurements to determine what the level of risk is.'

In one economic analysis [130], noise is considered an externality and the assumption is made that workers in noisy jobs are aware of the danger of hearing loss and therefore insist on higher wages.

To a certain extent it is true that workers insist on higher wages where there is danger to their health; this is called 'danger money'. In Britain at least it is not clear what the attitude of the unions is to danger money. Murray [131], in an official comment from the Trade Union Congress, implies two attitudes, one that where the employer faces insoluble technical problems in reducing a hazard he may ask workers to 'co-operate' with him, but what 'co-operate' means is not clear. The other attitude is that the Trade Union Congress leaves such decisions to its member unions and does not monitor this aspect of their activities. One would suggest that without involvement in the decisions and without monitoring there can hardly be a policy. In the building [132], construction [133], and civil engineering [134] industries in Britain, there is limited awareness of the risk of hearing damage by exposure to noise. In tunnelling it is seen as a source of discomfort, but not risk.

A short walkout by British Ford workers [135] on 22 August 1972 on account of bad working conditions, among which noise was specifically mentioned, must be one of the first instances of noise entering labour–management relations as a definite issue.

Other effects on health. The C–M group surveyed the literature on cardiovascular, adrenocortical, gastric and electroencephalographic effects, as well as the effects of noise on fatigue, and found no clear indication of damage. There are some indications in the work of Mohr *et al.* [136] and Ash'bel [137] that very low and very high frequency sounds can cause physiological damage. This suggests that the effects of these components in broadband noise should be investigated.

There have been some interesting investigations in Britain by Bryan [138] on the possibility of there being a connection between infrasound and car accidents, the infrasound being produced at motorway speeds in lorries and to a lesser extent in cars. The sound pressure levels produced at the dangerous low frequencies are not high enough to produce symptoms in all drivers but can produce symptoms in the minority of drivers who are more sensitive than average. The results so far are only suggestive and indicate the desirability of further investigation.

Are there any other effects? One's answer to the question depends on how one defines health. This author likes the World Health Organisation definition: 'Health is a state of complete physical, mental and social well-being, not merely an absence of disease and infirmity.' If one accepts that definition, one must say that disturbance to sleep, or for that matter, rest and relaxation, is a case of damage to health, whether it is caused by jet aircraft noise near airports, traffic noise penetrating into houses, or by the sonic boom.

Psychological damage. The problem of adjusting to hearing damage, and this does not mean only severe damage, but moderate damage too, has been well described by Sataloff [139] and by Ramsdell [140]. The description by Sataloff of the problems faced *off* the job by a man like a riveter who has hearing damage is very revealing:

'A hearing impairment may cause no handicap to a chipper or a riveter while he is at work. His deafness may even seem to be to his advantage, since the noise of his work is not so loud to him as it is to his fellow workers with normal hearing. Because there is little or no verbal communication in most jobs that produce intense noise, a hearing loss will not be made apparent by inability to understand complicated verbal directions. However, when such a workman returns to his family at night or goes on his vacation, the situation assumes a completely different perspective. He has trouble understanding what his wife is saying, especially if he is reading the paper, and his wife is talking while she is making noise in the kitchen. This kind of situation frequently leads at first to a mild dispute and later to serious family tension.

The wife accuses the husband of inattention, which he denies, while he complains in rebuttal that she mumbles. Actually, he eventually does become inattentive when he realises how frustrating and fatiguing it is to strain to hear. When the same individual tries to attend meetings, to visit friends, or to go to church services and finds he cannot hear what is going on, or is laughed at for giving an answer unrelated to the subject under discussion, he soon, but very reluctantly, realises that something is really wrong with him. He stops going to places where he feels pilloried by his handicap. He stops going to the cinema, the theatre or concerts, for the voices and music are not only far away, but frequently distorted. Little by little his whole family life may be undermined, and a cloud overhangs his future and that of his dependants.'

This may have important implications because ageing produces the same sort of hearing damage as noise exposure, as may be seen in Figure 10.5, and there may be some danger that an increasing

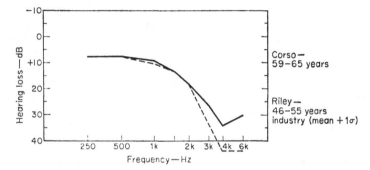

Figure 10.5 The similarity between hearing loss caused by ageing alone (presbycusis) and hearing loss caused by exposure to noise on the job. Middle-aged men who have worked in noisy jobs have *worse* hearing than old men who have *not* been exposed to noise

fraction of old people (in developing countries) may be having emotional difficulties due to hearing loss. The description of the riveter's problem may also explain the irritability that may come from their difficulties at home. In fact, Jansen [141] observed that these workers had fewer interpersonal difficulties at work than at home, just as Sataloff describes.

Some people active in noise abatement [142] make statements that noise drives people crazy, and as much as the C–M group might have wanted to confirm this, they could not honestly do so,* although a man living near Kennedy Airport wrote to the author that he couldn't have sex with his wife when aeroplanes were

---

* When the Wilson Report [143] was published in Britain in 1963 it was proposed that the incidence of mental illness was not noticeably higher around Heathrow Airport than elsewhere. However, more recently Abry–Wickroma *et al.* [144] have reported that there is some evidence of increased mental illness near Heathrow. In these questions it is notoriously difficult to draw unequivocal conclusions so the problem may be regarded as a serious one, but the population at risk is large.

going over. This might be due to a quite normal (*sic*) fear among those living very close to airports that the aeroplanes will crash on their houses. Indeed, this was found to be an important variable in the Tracor study [145]. Broadbent [146] clearly puts the case that when a large minority, estimated variously as 10–20 per cent, are very susceptible to being annoyed by noise they cannot be written off as neurotic, unreasonable, or not worth listening to. The C–M group did, however, ask the opinions of people who are working actively in the field and got the following replies:

'I am not aware of any work, recent or ancient, that has made a clear case for psychological damage caused by moderately high levels of noise as would cause hearing damage to only a small fraction of the people exposed. There has been a great deal of talk about this kind of thing and many opinions expressed publicly by psychiatrists, cardiologists, psychologists, and civilisation. I personally fear that over-emphasis on 'damage' may backfire when people come to realise that the truth of the matter seems to be simply that people don't like loud noise and don't like being disturbed.'

'My present feeling is that exposure to high intensity noise does not result in change of behaviour secondary to irreversible damage to the central nervous system. However, I would not venture a guess as to the interaction of personality variables with various forms of noise.'

'... I think we have a real hope of understanding what noise does in detail to the behaviour of people, and I want to pursue that. I would, however, add the general caveat that, if one could settle the question of the effects of a long continued background of noise upon the state or kind of well-being of people who had had an opportunity of becoming accustomed to it, this might potentially be very important in its results. However, I think the probability of reaching a definitive conclusion on this is extremely low and that the line of study is not, therefore, a very tempting one at this time.'

Clearly, even if there is only a small effect there are nevertheless so many people exposed that even a small effect must be reckoned as an important one.

There is a clear analogy with the problem of air and water pollution here; Herfindahl and Kneese [147] have pointed out that some 500,000 chemicals have been released for commercial use in the United States and that test programmes to find out the long-term effects cost something in the order of $100,000 each. Moreover, such test programmes are not unequivocal in assessing the level of risk. Thus, people are faced with the difficult problem of deciding whether to spend large amounts of money in the uncertain hope of finding out whether a serious long-term danger has been introduced into the environment. There have been calls for a moratorium on new technology [148] to give people time to assess the risk they are in. As the question of how risky a life people want to have is fundamentally a political one, scientists have nothing special to contribute, except to try to make clear the known levels of risk in the existing situation, so that people can make comparisons and decide for themselves.

Annoyance. Beside the physiological and psychological effects already mentioned, another major effect on people is annoyance. It is not immediately obvious whether traffic noise annoyance (more people) is greater than the annoyance caused by jet aircraft noise near airports (more severe annoyance). Again, there is a value judgement to be made; is it worse for a few people to suffer a lot than for many people to suffer a little? This is a typical case in which it is tempting to quantify the suffering in terms of loss in house values.

1. *Traffic noise.* From the data in Table 3, one can work out how much the various noise levels are above the ISO criterion.

2. *Jet aircraft noise.* One can easily work out the number of people annoyed by American airports by taking again the fifteen major air traffic hubs with airport close in and assuming that the percentage of people annoyed is 27 per cent, the same as it was at London Airport. Then 27 per cent of half the population of each city is assumed to be affected. When half the population is greater than 1,400,000, the number is purposely underestimated by taking only 378,000 people annoyed. This corresponds to the 27 per cent in the London Airport Study. The total number of people annoyed in the fifteen major air traffic hubs is then about 1·5 million.

Here one should remark that there is a common, and disturbing, tendency in the United States to identify complaints with annoyance. This is found in places as far removed as the mass media, for example the Sherrill article in the *New York Times* [149] and the professional literature, specifically the report of the Jet Aircraft Noise Panel for the Office of Science and Technology [150]. It is disturbing to find this tendency among professionals who should know better; one gets the impression that they never

Table 3. *Insulating residential units against traffic noise*

% Population		Noise levels inside residences due to traffic		Insulation required to meet ISO criteria (dB(A))	
Standard Metropolitan Statistical Areas	Min. night	Average dB(A)	Max. day	Day	Night
1	73·5	29	31	38	3	4
2	2·0	54	56	65	30	29
3	4·0	42	44	51	16	17
4	8·5	49	51	58	23	24
5	5·5	46	48	55	20	21
6	5·0	57	59	66	31	32
7	0·5	54	56	63	28	29
8	0·5	59	61	63	33	34
9	0·5	56	58	65	30	29

ISO inside noise levels (residential urban)	Day	Night
Single windows closed	35 dB(A)	25 dB(A)

read the London Airport Study [151] in which McKennell makes the point quite clearly.

In the London Airport Study, the complainants are viewed in two ways:

1. As representatives of the 11 per cent of the 1·4 million people in the twenty-mile area who are equally annoyed; or
2. As representative of the larger number, 27 per cent, of the 1·4 million who are 'annoyed', i.e. people who had an annoyance score of 3·5. These were the people 50 per cent of whom reported disturbance to rest and relaxation and being kept from going to sleep, and were somewhere between 'moderately annoyed' and 'very much annoyed'.

McKennell [152] makes further comments about the representativeness of the complainants as follows:

'Whether it is taken as 11 per cent or 27 per cent of the 1·4 million population, the minority in question is a very substantial one. When sensitivity is shown by such large numbers of people, it must be regarded as a normal phenomenon, in the sense that it can be expected to occur among similar large numbers in any ordinary population. It is the feelings of this substantial large uncomplaining yet sensitive minority, that the complainants make articulate.'

Now, can these estimates be made from another point of view? Let us recall that the 27 per cent figure from London Airport Study was accepted as typical.

The 27 per cent figure was based on an annoyance score of 3·5. At such a score, jet aircraft noise annoyance wakes up 50 per cent more people than other noise. Suppose one were to say instead that aircraft noise is annoying when it wakes up only (*sic*) 25 per cent more people than traffic noise, i.e. assume that it is unacceptable at this lower rate of annoyance. Then one would choose an annoyance score of 2·0 and the fraction annoyed goes up 27 per cent to 50 per cent. In this case the total number of people annoyed by Heathrow becomes 2·8 million. In other words, as the population of Great Britain is 55,000,000, this means that one person in every twenty in the entire country is annoyed by just *one* airport, and this leaves out inter-continental airports like Manchester, Glasgow–Edinburgh, and Gatwick.

Sonic boom. The remaining major source of annoyance (in the United States) is the sonic boom. The C–M group made the following findings:

The laboratory tests by Kryter and Pearsons [153] and by Broadbent and Robinson [154] and the full-scale tests in St Louis [155], Oklahoma City [156], and at Edwards Air Force Base [157] are well known. Perhaps less well known are the following details.

1. In Kryter and Pearsons' study [158] half of the subjects said they could not learn to accept the simulated booms, although the booms were produced at the subjects' will. This means that they had some chance to prepare themselves.

2. In the St Louis study, there were a number of details that were not clear and the C–M group were unable to obtain clarification of them even after extensive correspondence.*

* Letters written directly to authors of NASA reports were not answered by them but by their supervisors, perhaps to guard against release of information that NASA think the public ought not to have. This is a common problem in what might be called preventive technology, by analogy with preventive medicine. In this author's experience [159] and in the experience of other investigators [160] it is extremely difficult to extract information from industry. This is even true of nationalised industry in Britain; British Rail will not yet say what their safety policy is for the Advanced Passenger Train [161].

The number of complaints rose sharply at the end of the test series, when a local newspaper published an editorial against the tests. As this seemed likely to happen if overland SST flights took place, they asked about its significance. This explosive rise in complaints is well described by the US Air Force officials involved in the test series. Their description is lengthy but is included as an Appendix because it has so much relevance to the question of making decisions that may result in disruption of the environment.

These were claims from 'unpopulated' areas at the edge of the city. The C–M group asked how many of these claims were allowed. The flight path was chosen to the east of the city; there is a rich residential area to the west. The group asked whether it was thought that such information could be used to predict how much people in a richer residential area complain. There was no information about the complaints from the night-time booms. As these will obviously cause more annoyance, the group asked for details.

The original NASA report [162] showed a larger total of 'valid incidents'* than the published summary article [163]. The group asked about this.

3. In the Oklahoma City Study [164] the percent annoyed somewhere between 'seriously' and 'more than a little' increased to 56 per cent at the end of the series. And Borsky estimates the potential complaints at between 25 and 75 per cent—25 to 75 per cent of 50 million people if Kryter [165] is right! The booms in Oklahoma City were all produced during the day, yet 18 per cent of the people interviewed were awakened: This may seem startling; one might wonder how many could have been asleep during the day, but some of the booms were at 7 o'clock in the morning. Using the detailed records of when the booms occurred [166] and records of electricity demand [167], one can estimate the fraction asleep as 40 per cent. Of these 18 per cent (of 40 per cent) were awakened, or 45 per cent of people sleeping. This is comparable with the findings of Lukas *et al.* [168] which indicates that whilst children are not awakened by sonic booms at all, middle-aged and old men are, more than 50 per cent of middle-aged men and more than 25 per cent of old men.

The maximum peak overpressure in the Oklahoma City tests was 1·5 lb per square foot instead of the 2·0 lb per square foot that was originally intended. This poses a problem of extrapola-

* A valid incident is recorded when a claim for damage is substantiated by engineering examination.

tion. Moreover, as is pointed out in the discussion of the Edwards Air Force Base tests, these were B-58 booms, so there is an additional problem of extrapolation. Added to this, the results of the St Louis, Oklahoma City, and south coast of England tests [169] are very different as may be seen in Figure 10.6; so different as to raise serious problems of prediction. Which trend is one to believe? This has been discussed by Wakstein [170].

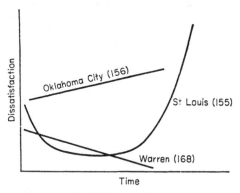

Figure 10.6   Changing dissatisfaction with time in sonic boom tests

In the Edwards Air Force Base Tests [171] people were warned just before each boom! This faces planners with an additional problem of extrapolating the element of surprise.

Some of the main conclusions are given below:

1. (a) When indoors, subjects from Edwards Air Force Base judged booms from the B-58 at 1·69 psf nominal peak overpressure outdoors to be as acceptable as the noise from a subsonic jet at an intensity of 109 PNdB measured outdoors.

   (b) When indoors, subjects from the towns of Fontana and Redlands judged the boom from the B-58 at 1·69 psf nominal peak overpressure outdoors to be as acceptable as the noise from a subsonic jet at an intensity of 118 to 119 PNdB measured outdoors.

   (d) When indoors, 27 per cent of the subjects from Edwards and 40 per cent of the subjects from Fontana and Redlands, combined, rated the B-58 booms of nominal peak overpressure of 1·69 psf as being between less than 'just acceptable' to 'unacceptable'.

2. (a) When of approximately equal nominal or measured peak overpressure and when heard indoors and judged against

the aircraft noise, the boom from the XB-70 was slightly less acceptable than the booms from the F-104 or B-58 aircraft. When heard outdoors and judged against aircraft noise, the boom from the B-58 was slightly less acceptable than the booms from the SB-70 and F-104 aircraft.

The most significant ones are 2(a) and 1(b). 2(a) implies that the estimate of annoyance for St Louis and Oklahoma City is too low and 1(b) is important because 118 PNdB is unacceptable in any circumstances. Even at Kennedy Airport the New York Port Authority limit is 112 PNdB [172]. Although it might be argued that as the official limit is in fact often exceeded [173], and as there has not been massive community action against Kennedy Airport, perhaps in some sense a level higher than 112 PNdB is acceptable. However, even with the complaint level as it is, one of the runways has been lengthened [174] solely to reduce annoyance, so it is not likely that a higher level would be acceptable. Remember, also, the sexual difficulties of one man who wrote to complain.

Recently there have been supersonic overflights of the Concorde along a north–south route across the West of England and parts of Scotland.

Although the great majority of Concorde flight routes would by admission of the British Government have to be oversea routes because of the disturbance caused by sonic booms, it has recently been suggested [175] that certain overland routes involving 'sparsely populated' areas be considered as potentially acceptable. These proposals are most interesting from the standpoint of the value judgements mentioned earlier; there seems to be a fundamental inequity in subjecting people in the wilderness to sonic booms but not people in or near cities. Moreover, who will decide what 'sparsely populated' means, and how?

(i) *The cost of annoyance.* The cost of annoyance has been commonly translated into loss of house values, but the validity of such an approach depends on whether the seller sells because of not being able to live with the noise, and whether the houses from which the buyer chooses are being sold because the sellers cannot live with the noise or for some other reason. However, Crockett [176] has reported that traffic noise (vibration transmitted through the air, not the ground) can cause large vibrations in structures like churches. The damage caused by these vibrations is long-term and has not yet been put in monetary terms.

For that matter, economic indicators like GNP do not necessarily reflect the damage done by pollutants. Doctors, at least in the States, are not underselling their services in treating people with hearing damage and it is not obvious that the additional GNP created by their services is outweighed by the loss in productivity if indeed there is any. Insurance companies are not losing money on Workmen's Compensation insurance so they too are adding to the GNP. Therefore, one might well ask if hearing damage is costing the United States (or any country) anything. One might even argue that it is no economic loss for lots of people to be suffering some degradation in health, say hearing, as long as it does not affect productivity. There is a conflict between a greater GNP and public health, in this case, less noise-induced hearing damage.

(ii) *Structural damage.* In the case of subsonic jet aircraft noise, it is well known [177] that houses near airports vibrate and shake when jets take off. The C–M group were unable to find any evidence in the literature of damage caused by this vibration.

However, the sonic booms produced by the SST do cause structural damage, mostly to plaster and windows. Similarly Crockett [178], has pointed out that damage to stained glass will be accelerated by the bowing in and out of the windows as they respond to sonic booms. This will gradually wedge open the lead holding the individual piece of glass and allow oxidation and other deterioration to take place more quickly.

Data of damage to houses from the St Louis, Oklahoma City and Chicago sonic boom tests [179] may be used to estimate the damage in the States that could be caused by future SST overland flights. The relevant data on routes are abstracted from the IDA report [180].

First the routes classified by the IDA as potentially feasible* were considered. This gave about 24,500 miles of routes, leaving out 200 miles' flow subsonically at each end of a flight; and following a suggestion of Maxwell [181] to take a fifty-mile wide carpet for the boom, a figure of say 1.2×10^6 sq. miles subjected to sonic booms was obtained. The area of the United States is 3.6×10^6 sq. miles. This gives about one-third of the 50 million rural population of the United States subjected to sonic booms, or 16.7 million people. This is smaller than Kryter's estimate [182] and if Kryter is right, this confirms an earlier suspicion that the edges of cities and some entire cities will be subjected to booms as well.

One might ask the question: How much do the airlines need the overland routes? From the Institute for Defence Analyses data

* The domestic routes classified as feasible are overwater routes.

there are $12 \cdot 24 \times 10^9$ seat miles per year of feasible domestic overland SST flights. This amounts to about 500×10^6 per year. Thus, there appears to be a strong incentive to fly the potentially feasible routes. On these grounds, it is assumed that there will be strong pressure for the potentially feasible overland flights to take place. Then Kryter's estimate [183] of the number of people subject to booms may be combined with the St Louis, Oklahoma City and Chicago damage figures [184] to estimate the structural damage. We have $1 \cdot 53 + 0 \cdot 83 + 0 \cdot 55 = 2 \cdot 72 \times 2$ for a two-sided path $= 5 \cdot 44$ valid incidents per flight per million people. Taking $116 as the average claim [185], one has $5 \cdot 44 \times 50 \times 15 \times 116$ dollars per day $= \$473,000$ per day $\times 360 = \$170$ million per year.

The problem of making such an estimate is complicated by the fact that very intense booms are statistically unlikely but nevertheless possible as Lundberg [186] has pointed out and that such booms will cause very severe damage. Clarkson and Mayes [187] discuss the problem but as may be seen in Figure 10.7 the contri-

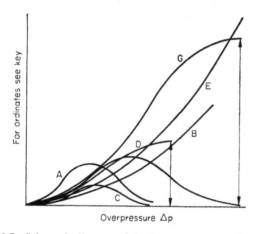

Figure 10.7   Schematic diagram of the damage caused by sonic booms

Key
A = probability of occurrence of a boom of overpressure Δp
B = damage caused by a boom of overpressure Δp—'square law'
C = statistically expected damage A × B
D = cumulative expected damage
E = as B but with 'third power' damage
F = as C but with 'third power' damage
G = as D but with 'third power' damage

Note that the majority of the damage in G is caused by the extremely unlikely severe booms which contribute negligibly to the damage in D.

bution to the total damage from such booms can in principle be very significant. However, what is not known is whether such booms occurred in the small (*sic*) sample of booms produced in the Oklahoma City tests. Thus, one cannot be sure that the estimates made are not gross underestimates. Moreover a more recent estimate by Kryter [188] based on routes and population densities gives a boom-person exposure more than double the boom-person exposure of his earlier estimate. Indeed, according to this estimate, some three million people would hear *at least* thirty-five booms a day. Apparently no such estimate has been done of the structural damage that a fleet of Concordes would inflict on Britain.

Besides considering such estimates of possible damage, one may also consider one of the other decisions that had to be made in the case of the American SST, namely the evaluation of how much more airlines' passengers would pay to travel faster. The American government commissioned the Institute of Defence Analyses to carry out a study [189] on this question as part of an overall economic analysis of the economic feasibility to the SST. In the IDA analysis, various major airlines were asked to give their best estimates of the surcharge as a function of time saved. Typical estimates of how many passengers would prefer to fly SST as a function of how much extra it would cost are shown in Figure 10.8.

The IDA decided to make the most cautious estimates of people's willingness to pay extra to fly faster and therefore based their calculations on people being willing to pay only an amount equal to their earnings rate for the time saved. This was less optimistic than basing the calculations on the average of the functions submitted by eight international and four domestic airlines. As the IDA put it in their report the more optimistic assumption using the average would be 'slightly more favourable to the SST' than the straight earnings model. This resulted in the prognosis being doubtful.

The FAA [190] proposed a function based on one and a half times the earnings rate on the grounds that the earnings rate model would be unfavourable to business travellers and that the FAA function was a compromise between the earnings rate model and the airline average (which was equivalent to twice the earnings rate). But when one examines the actual figures that went *into* the airline average it may be seen that it is twice the earnings rate *only* because two of the eight international airlines and two of the four domestic airlines made *very* high estimates of what people would be willing to pay. At one representative point

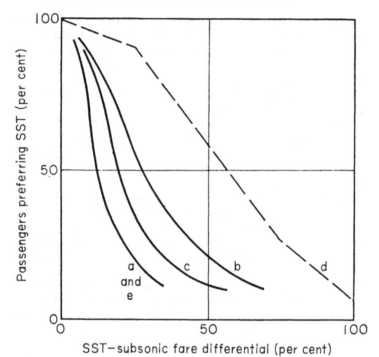

Figure 10.8   The FAA 'Compromise'

a = IDA earnings rate model
b = airlines' average estimate
c = FAA compromise: halfway between a and b
d = one of the few very high estimates that accounts for the airlines' average
e = airlines' average if the very high estimates are omitted

on the curve these airlines' estimates were three to six times the IDA estimates and in one case *twelve* times the estimate of another airline. These airlines making very high estimates were the smaller airlines and therefore presumably less experienced and less able to support the data gathering and processing required for intelligent prediction. Indeed if these very high estimates are *omitted* the airlines average is almost exactly equal to the IDA earnings function and the FAA function is then 50 per cent optimistic.

This is a common statistical problem, that of an average disproportionately influenced by a few very high values. This procedure led to the conclusion that the SST was economically feasible.

To take the most charitable interpretation the FAA honestly

believed that the very high estimates of the small airlines should be given equal weight. The least charitable interpretation is obvious, and if one accepted it one would be tempted to ask what the FAA would have done if even with their function the prognosis had been dubious.

In saying all this, which at first glance seems to relate only to the SST, the more general point should be grasped that the decisions that affect disruption of the environment are difficult decisions to make not only because they involve uncertainties, but also because they have such large consequences. It is the view of this author that notwithstanding the cogent and reasonable arguments of Bayesian decision theorists like Raiffa [191] one's estimate of the uncertainties *is* affected by one's view of the desirability of the consequences which in itself is a value judgement. As it is likely that this will lead to biased decisions being made, one might argue that the proper way for such decisions to be made is not by supposedly 'objective' professionals, but by people and professionals participating in an informed way.

Unfortunately, the rationale behind the decisions to go ahead with the Concorde is not so openly available [192]. We argue here for laws requiring that much more information be made available to the public.

The Cost of Quiet

Houses. The C–M group estimated the cost to protect all new houses built each year in standard metropolitan statistical areas in the United States; new houses because their acoustical condition is known at the time they are finished. The acoustical condition of older houses varies widely and is not easy to predict from a knowledge of their construction. It should be pointed out that the cost of protection by providing insulation in the design stage, rather than *after* the house is built, is significantly less; it has been estimated at as little as 5 per cent of the uninsulated house cost [193].

An estimate of the cost of insulating against traffic noise all the new houses built each year in the States has been obtained by starting with the Boston traffic study mentioned above. The main calculations are given in Table 4. The traffic flows in cars per day have been converted to noise levels by use of the BBN empirical expression.

The maximum traffic in the heavier of the two rush hours was calculated by assuming that the Pittsburgh Golden Triangle traffic data [194] were applicable to Boston. The traffic between 11 p.m. and midnight was calculated from these data. The noise levels inside the houses recommended by the ISO [195] are given as 25 dB(A) at night and 35 dB(A) during the day, whereas the 10 per cent levels in the Wilson report are 10 dB(A) higher. The houses are assumed to be 'residential urban' with single windows that are closed.

In planning studies these recommended levels are compared with levels extrapolated from levels exceeded a certain fraction of the time, usually 10 per cent measured with the expected traffic flow, and at a certain distance from the centre of the nearside lane, usually twenty-five feet. The levels based on GLC data [196] apply to high traffic flows, more than 40,000 cars a day, and are 75 dB(A) during the night and 83 dB(A) during the day.

The figures in Table 4 show the cost of additional insulation against noise; with these data the average cost per house works out to be $1,030. The total cost is thus 10^6 new houses per year [197] at $1,030 per house = 10^9 per year. For comparison all we can say from British experience [198] is that double glazing with mechanical ventilation costing £500 per dwelling can give 40 dB(A) reduction and double glazing with staggered openings costing £300 per dwelling can give 20 dB(A) reduction. Even at five dollars to the pound it is still cheaper to do this job in Britain.

Table 4. *Cost of additional noise insulation for houses in the US*

Additional insulation dB(A)	Cost per dwelling unit
0–5	$200
5–10	500
10–15	1,000
15–20	1,750
20–25	3,000
25–	5,000

The C–M group considered this a very large figure, and certainly larger than the cost of better mufflers for lorries and buses. (The number of large lorries and buses in the US is 1·2 million [199].) The additional cost of a better muffler is given as $20 to $30 [200]; these mufflers must be replaced, say, once a year. Thus the cost of traffic noise reduction by mufflers on lorries and buses varies between $24 and $36 × 10^6 per year. However, direct comparison between the two figures must be approached with caution without consideration of the overall noise reduction that could be brought about by the mufflers. Furthermore, even if the mufflers reduced the exhaust noise by a large amount, there would still be the tyre noise; at present we do not have data for truck tyre noise in traffic.

This is a simplistic conception of the problem; as in the first place the reduction that can be achieved by improved silencers [201] is only 3 to 4 dB(A) [202]. From Table 5 the reductions that are needed in commercial vehicles in urban areas are more like 20 dB(A) than a few dB(A). And as soon as one starts talking about reductions of this magnitude, one is talking about much higher costs, of the order of thousands of dollars according to data reported by Cummins Engine Co. [203].

Table 5. *Insulating residential units against traffic noise (US)*

Standard Metropolitan Statistical Areas	% Population	Min. night	Noise levels inside residences due to traffic		Insulation required to meet ISO criteria (dB(A))	
			Average dB(A)	Max. day	Day	Night
1	73·5	29	31	38	3	4
2	2·0	54	56	65	30	29
3	4·0	42	44	51	16	17
4	8·5	49	51	58	23	24
5	5·5	46	48	55	20	21
6	5·0	57	59	66	31	32
7	0·5	54	56	63	28	29
8	0·5	59	61	63	33	34
9	0·5	56	58	65	30	29

ISO inside noise levels (residential urban)	Day	Night
Single windows closed	35 dB(A)	25 dB(A)

The C–M group considered that engine and drive-train silencing could only bring about limited changes because of the problem of tyre noise, but Alexandre [204] properly pointed out that

'. . . it should be remembered that under normal urban conditions tyre noise (or rather the noise from tyres interacting with the road) is not significant since it becomes predominant only at high speeds, i.e. on motorways and on roads outside densely built-up areas. In urban centres, the effect of reducing tyre noise will be felt only in the long run, after other sources of vehicle noise have been significantly reduced.'

The nature of the problem can be clearly seen in Figure 10.9 [205]. At the lower speeds that would be found in urban areas where large numbers of people would be exposed to commercial vehicle noise the margin that has to be made up before one would begin to worry about tyre noise is very great indeed; at 15 mph in second gear there are 26 dB(A) to be made up. Only at 30 mph in fourth gear is the margin small, about 5 dB(A), and this is for constant speed. Acceleration increases the noise level by 10 dB(A) [206].

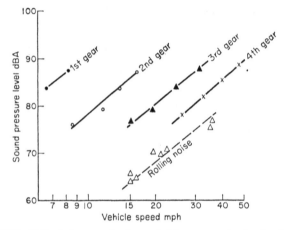

Figure 10.9   Fully laden constant speed drive-past noise levels of a diesel lorry [205]

Machines.   It is easy to estimate the cost to quieten machines in American industry if one has some actual cost data to start with. The C–M group were unable to find any in the open literature, but were very fortunate to get some from Mr G. E. Parsons of Hercules Inc. [207]. These data cover costs of noise control over a period of thirteen years; the average cost works out to $70 per dB per worker. Other sources quote different values; the Committee on Environmental Quality [208] gives a figure, expressed in similar units, of $26 per dB per worker.

There are $5 \cdot 5 \times 10^6$ workers in metalworking [209]; there are $16 \cdot 2 \times 10^6$ workers in manufacturing as a whole [210]; this includes production workers and people in administration.

Let us say that machines need to be quietened in between 10 per cent and 100 per cent of cases and that the amount of noise reduction required is between 10 dB and 30 dB. Then taking minimum figures we have 10 per cent $\times$ $5 \cdot 5 \times 10^6$ workers $\times$

10 dB $\times$ \$70 = \$385 $\times$ 10^6, and taking the corresponding maximum figures we have 100 per cent $\times$ 16·2 $\times$ 10^6 workers $\times$ 30 dB $\times$ \$70 = \$3·4 $\times$ 10^{10}.

These figures are for improving all the machines that exist now; assuming that machines get replaced every ten years, at most \$3·4 $\times$ 10^9 per year need be spent.

The role of hearing protectors. The corresponding cost for ear muffs (at \$5 per pair, and capable of reducing noise by as much as 40 dB) would range from \$27·5 million to \$89·1 million per year.

One recognises that one may not have included all machines; after all, it has already been estimated that there are 34 million workers in noisy jobs. The rough estimates thus indicate that the cost of modifying the machines is much greater than the cost of earmuffs.*

Then the sensible approach is the one which enlightened industry is taking [211], give everyone earmuffs, but quieten at least the noisiest machines too; because the workers will not always wear their earmuffs properly. Providing earmuffs is acceptable as a control measure only if it is monitored to see whether it is working [212]. This raises the question treated by Delany [213] at the 1970 NPL conference (and also discussed by others), of whether the gradual changes in hearing levels that are revealed by serial, i.e. periodic audiometry are real changes or not. The situation is uncertain; earmuffs may be supplied but one does not know whether workers wear them properly so they may be accumu-

---

* The estimates are based on averages, so must include some cases where machine quieting is very expensive. One learned of a case [203] where \$10,000 was spent to quieten a large press something like 25 dB. However, there are obviously other cases where machine modification may well be cheaper than using earmuffs. Further investigation of this is clearly justified.

In the discussion at the 1970 NPL conference, it was this author's strong impression that, probably because of cases like this, participants from industry had found it impossible to believe that noise reduction could be bought cheaply. Because of this the fact that even the highest estimate of the total yearly cost for quietening of machines is only about $\frac{1}{2}$ of 1 per cent of (the American) GNP [214] tended to be disregarded. The conclusion to be drawn from this estimate doesn't *have* to be that the noisy industries must carry the largest share of the cost of noise reduction. This is the case at present in the States on account of the Walsh–Healey Act which requires that companies with Federal contracts must meet certain levels, 90 dB(A). Another possibility is that a uniform levy per worker be imposed on all industries just as it is in the case of social security. However, this didn't even get discussed, perhaps because of the fears in British industry that they may be made to account for the hearing damage they have caused over the years. Indeed, the jute mills in Dundee have not changed their noise levels for as much as seventy years [215]!

lating hearing damage but one cannot tell unambiguously until there has been enough to measure, say 10 dB average at the usual AAOO frequencies. This could mean a change from mild to serious hearing loss.

Conclusions

What can be done? People can be educated to the danger. They can be educated to protect their ears. We can tell our children not to make bangs in each other's ears with the same urgency that we tell them not to poke in each other's eyes. People can be educated to wear earmuffs when doing any noisy job or when they are in a noisy environment.

It has been suggested [216] that people who are considering working in noise could make an informed decision if only they were clearly informed of the nature of the risk to their hearing. This is an important suggestion in a democratic sense because it goes in the direction of people having an influence on the decisions that affect them. This implies communicating to them in a clear way what it would be *like* to have hearing damage. Filtered recordings of a conversation have been available for a long time and with improvement could still be used to give a rough indication of the effects, but some people might not be willing to let their hearing go that far. A way is needed to communicate to people what would be lost, but before looking for ways to communicate this, a clearer idea of what could be lost is needed. There is a need for a long list of sounds like clocks ticking, bacon frying, leaves rustling, wind in trees, rain, combing one's hair, skis on snow, etc.

Machines can be muffled. This is expensive, but new machines could be more easily designed from the start to be quiet. Federal laws might be passed to require quiet machines. The role of the Federal government has not traditionally been to set levels for things like noise; this has been left to the states. However, the Federal government has acted through the Federal Aviation Administration in the case of annoyance due to airport noise and could conceivably act in interstate commerce to protect against hearing loss. From the exchange between van Atta and Wakstein at the 1970 NPL conference [217], there is apparently some hope that in the United States at least the Federal government will be able to regulate noise, however, the matter of enforcement is a difficult one. As van Atta pointed out [218] the existence of a law is no guarantee that it will be adequately enforced. The situation in many ways resembles that in Britain where the Factory Inspectorate is badly understaffed [219] and often cheated [220] and the fines are trivial [221].

Value judgements. Perhaps the most important is the observation that it is not possible to find just one number or one set of numbers to describe the seriousness of the problem.

Conclusions about the seriousness of the problem depend very strongly on certain *choices* such as:

(a) Do we* use an annoyance score of 3·5 or 2·0 in defining airport noise annoyance?

(b) Do we use the Composite Noise Rating [222] or the Traffic Noise Index of Langdon and Scholes [223] or the 10 per cent noise level L_{10} [224] *or* the Noise Pollution Level proposed by Robinson [225]?

(c) Do we use a high or a low value of Workmen's Compensation awards for noise-induced hearing loss?

(d) Do we use the AAOO or the Australian method or the British Hygiene Occupational Society standard to assess hearing damage?

(e) Do we take quality of the environment into account in defining hearing damage or not?

(f) Do we use the lowest or the highest value for the annoyance from the sonic boom?

(g) Do we follow the American example [226] and set 90 dB(A) as a dangerous noise level for long-term exposure or follow the Dutch example and choose 80 dB(A) [227]? In the end this amounts to deciding (1) how large or small a minority will suffer (2) how much hearing damage, (3) measured *a certain way.*†

All of these decisions are value judgements.

There is increasing acceptance of the view of Heymann [228] and Blum [229] that such decisions are political decisions and must be made by the public. Perhaps even more important there is increasing acceptance of the views of Churchman [230] that the public must be well informed.

Another trend that is likely to take longer to bear fruit is that of *real* participation in the formulation of alternative plans to deal with, say, pollution. The history of such attempts at participation in highway planning is dismal. The Liverpool M62 [231] is an example of a failure in public participation. But one continues to hope.

* We = professionals *and* people working together.
† This is called a Damage Risk Criterion (DRC).

Whatever choices are made, they are bound to be criticised in such crude terms as 'idealistic' or 'reactionary'.

It may be said that it is not up to the engineer or the economist to express an opinion about which way the choice ought to be slanted, but this author agrees with Ackoff [232] that professionals are obliged to express an opinion about this and about how *people* (not planners) ought to use their information. This author thinks the slant should be in the direction of allowing only a minority to suffer any degradation in health, taking the same definition of health proposed by the World Health Organisation. If one were to choose the 'idealistic' estimates one would come up with a recommendation that, say, in the States $1 to $10 billion a year be spent on noise control, but even 'idealists' might have trouble justifying that amount of money with all the other demands on the resources of the United States. What is a reasonable balance?

Whether one is an 'idealist' or a 'reactionary' it seems that some way is needed to arrive at a recommendation so that it could be related in the Congress or the Parliament, etc., as the case may be. One suspects that these balances are arrived at in practical political terms by the 'squeaky wheel getting the most grease', and one recognises that whatever is proposed will get modified, but something has to be proposed to start with. One would like to be considerate to those people who are sensitive to noise and one also has some idea of what might be called the good life where there would be relative freedom from noise, but there are other demands on available resources.* How does one make up one's own mind among all these conflicting demands? Rather than take the narrow view that all will be solved by 'going interdisciplinary' and having teams of professionals work on these problems, a perhaps more constructive and clearly more democratic approach is for professionals to work side by side with the people affected from the very start.

Further Studies Needed

A study is needed of the noise exposure experienced by people riding in cars, lorries, buses, subways, trains and motorcycles. Special emphasis should be given to cars, buses and subways on

---

* This is not to say that all problems are solved by spending money, indeed, one of the most commonsense things suggested in the Wilson Report is that people might just be a little more considerate and make less noise. The hallmark of the good designer is the ability *not* to spend money when a solution can be achieved without spending money.

account of the danger of hearing loss. This study could be carried out by means of portable recording equipment. The C–M group discovered for instance, that noise levels in some American subways [233, 234] are high enough to suggest that rides twice a day with even nine hours in between will cause hearing damage in a minority of people.

A preliminary study of the increased annoyance due to greater usage of medium-range jets and business jets is needed.

Further study is required of the trends in Workmen's Compensation in states and countries that do not yet award for noise-induced hearing loss. Coming events in Britain, as compensation becomes widespread, should be particularly interesting. Another question might be formulated as follows: do the states and countries with low Workmen's Compensation awards for noise-induced hearing loss tend to spend less money on noise control and how does this vary by industry? On account of the co-operative attitude of one company in the United States, one is encouraged to assume that other companies will also co-operate in such a study and one recommends that it be carried out.

Further study is needed of methods of assessing impairment in hearing because even the method proposed by Atherley and Noble [235] involves value judgements.

Further study is needed of the loss of efficiency due to noise, in spite of the methodological difficulties [236] of such studies.

Further study is needed of the cost to quieten machines.

As the level of awareness of the risk of hearing damage in noisy jobs is an important factor in the economic analysis, a study of awareness is needed, to be followed by a campaign to increase it.

Discussion by people and professionals is needed to determine how well Workmen's Compensation awards for hearing loss represent the loss to society or the economy, to reconcile the variation in state award schedules, and to work out a way of deciding together what a reasonable award schedule or programme might be. One might also include in this study a comparison of awards for hearing loss with awards for other disabilities; one would hope to discover anomalies, to explain the reasons for these anomalies and to work out ways of deciding together what changes might be reasonable.

There is a need for a comparative study of the many different damage risk criteria, especially the recent one of Kryter *et al.* [239]. The value judgements implicit in them ought to be made clear, as should the reasoning leading up to the choice that was made.

Machines of all kinds have been getting more powerful over the years. Just in terms of the growth in electricity demand per capita in industrialised countries [238] this is clear. However, no specific information about increased power was available from the Machine Tool Manufacturers' Association [239] in Britain. A study is needed to determine whether techniques for noise control have kept up with this trend and what the future implications might be.

The most recent large-scale study of noise in industry was by Karplus and Bonvallet [240] in 1953. The study needs to be updated and this updating ought to be continued, say every ten years.

Further study is needed on what the long-term effects on the human race might be if a majority of people had hearing damage.

Millions of people have hearing damage that interferes with comprehension of speech. This sort of hearing damage is not satisfactorily corrected by hearing aids in their present stage of development, and the problems of making such hearing aids is a very difficult one. Nevertheless, further study aimed at developing adequate hearing aids is needed.

A study is needed to determine the subjective effects of noise induced hearing loss in its early stages, because it is not clear that the threshold value for interference with understanding of conversation is low enough, or that interference with understanding of conversation is a liberal enough criterion.

A public service advertising campaign on noise induced hearing loss is needed to increase awareness. Such a campaign would cost about $80,000 in the States and might be sponsored by the American Academy of Ophthalmology and Otolaryngology, the Industrial Hygiene Foundation, or the US Public Service. There are obvious analogies for other countries.

A publicly available study is needed of the formation of groups against noise and pollution in general. In the long run governments and industry respond to public pressure and such pressure groups can learn from each other's experience. A start could be made with the Air Pollution Control League of Greater Cincinnati [241].

A clearing house to make information about industrial noise readily available is needed. This could be sponsored by Government Health Departments, or an international one could be set up.

A statistical study is needed of the levels of background noise against which speech takes place. The data could be used to make speech audiometry more realistic.

A study is needed to develop damage risk criteria for summer jobs, seasonal work, and weekend exposure. Because of the important work by Burns and Robinson [243] on the concept of emission and the work of Baughn [243] on the combined effects of pathology and noise exposure it is now possible to predict hearing damage from exposure that is *not* the usual eight hours a day, five days a week, fifty weeks a year; this information should be made publicly available.

Conventional noise meters cost at least £150 each in Britain. It seems reasonable to assume that this puts small industry off buying them and that there is a need, and therefore a market, for cheaper meters. The Noise Abatement Society in Britain have developed one which sells for £10 [244], the Radio Corporation of America are now marketing a dB(A) meter that sells for 75 dollars [245] and Meyerstein [246] has developed a simple one, using integrated circuits, that offers hope of costing even less. More work should be done on these.

A study is needed of the emotional effects of presbycousis in an ageing population like that of the United States. Part of this would be a study of the detailed exposure of people to noise in environments not normally considered noisy.

Appendix

Events leading up to the Explosive Rise in Complaints in the St Louis Sonic Boom Tests [247]

Events in the St Louis area subsequent to January 1962 are of particular significance to the interpretation of the survey findings. A brief evaluation will be suggested and, then, a detailed account of events made by Lt Col. Almon A. Tucker, Staff Judge Advocate at Scott Air Force Base, will be presented.

There were nineteen confirmed booms in November 1961 while the initial interview took place, of which thirteen were special test flights concentrated within a week's time, and eight of which occurred about 11 p.m. or later. During December there was a sharp drop in booms to only six. In January prior to the second wave of interviews, there were only four test flights, all before 11 p.m. This sharp fall-off in boom exposure will be discussed in a later section in connection with a discussion of second-wave interviews. During the rest of January there were twenty-one additional booms (total of twenty-five), and during February there were fifteen. During March, there were twenty-two booms and during April there were twenty, of which fifteen occurred within the five days 2–6 April 1962. In addition to the unusually heavy concentration of booms during this period, it is believed the intensity of some of the booms was unusually severe due to unusual operational manoeuvres by the particular aircraft.

The observations of Lt Col. Tucker about a 'cumulative saturation point' may have some merit and should be further investigated. Certainly the importance of frequency and time of boom should be further studied. It should also be recalled that less than half of all respondents were convinced that others believed the booms were absolutely necessary, suggesting their own doubts about the unavoidability of the booms. The fact that a low complaint level may suddenly increase sharply if sponsored by local authorities and organisations has already been discussed. If the full potential of 10–27 per cent were to contact the Air Base, on the basis of 500,000 families, the total number of calls could have been 50,000–130,000. Obviously, the announcement on 6 April that flights were being discontinued prevented the potential build-up in complaints. The entire experience underscores the danger of relying on unanalysed overall complaint levels.

Lt Col. Tucker's report follows:

General public reaction to sonic boom phenomena was one of tolerance and forbearance through the period 5 July 1961–31 January 1962. Complaints and claims were received, of course, but the tenor of the letters generally was that of acceptance and understanding of the need for the training exercise. Since the first part of February, however, letters and telephone calls from complainants become noticeably more hostile and irate. The questions were invariably pressed, 'How long are these booms to continue?' and 'Why can't the runs be made somewhere else?' This rather smouldering public resentment was seized upon by the newspapers, especially the Editor of the *Globe-Democrat*, the only morning newspaper in St Louis, and utilised as the backdrop for a series of articles which appeared in the paper during the period of time that the continuation of the booms became intolerable to the general populace in the vicinity of St Louis, the following chronology of sonic boom events is given in detail:

Monday, 2 April

On Monday, there were three booms of moderate intensity at the following times and altitudes:

Time	Altitude
6.45 p.m.	45,600 feet
7.48 p.m.	42,000 feet
11.52 p.m.	45,000 feet

Tuesday, 3 April

On Tuesday evening, there were three booms of very great intensity. The night switchboard operators were deluged with calls of complaint, and one man insisted on speaking with the Base Commander. Instead, he was referred to Lieutenant Callahan, Assistant Base Claims Officer, of whom he demanded that the flights be stopped. This man was articulate, sober, and not the typical crank complainant, yet he was in a state of what may be termed frustrated indignity. This case is considered significant because later in the week this attitude was very prevalent in many complainants who were obviously in a state of great emotional disturbance and almost tearfully were demanding that the flights be stopped. The complainant on this night stated his intention of calling General Powers at SAC Headquarters or Bunker Hill Air Force Base, as soon as he finished talking to Lieutenant Callahan. The times and altitudes of the booms this day were as follows:

Time	Altitude
9.50 p.m.	48,000 feet
10.40 p.m.	47,300 feet
11.04 p.m.	48,000 feet

Wednesday, 4 April

The morning radio newscasts carried bulletins regarding a wall which had collapsed as a result of a sonic boom. A team of investigators from Scott visited the scene of the damage early in the morning and discovered that a section of weakened wall had fallen and that four witnesses who were playing cards in the house when the boom occurred, actually felt and observed the collapse of the wall. The evening newspaper carried a front page story about the wall collapse, and evening television newscasts carried pictures of it. Two additional booms occurred this day in the early evening, as follows:

Time	Altitude
4.32 p.m.	46,000 feet
6.46 p.m.	46,000 feet

Thursday, 5 April

On Thursday, the morning newspaper carried a front page picture and history of the collapse of the wall and various other items of alleged sonic boom damage in the St Louis area. In addition, an editorial appeared in which the *Globe-Democrat* retreated from its earlier stand that the booms were necessary and should be borne patiently by the populace. The paper now said that the flights were being overdone. Over seventy-five telephone complaints were received this day at the Claims Office, Scott, including calls from three almost hysterical people. One woman claimed that the aircraft creating a boom nearly struck her roof. Another woman stated that her sick husband was on the verge of hysteria, and a minister stated that he was flying to Washington immediately to see that the flights were stopped. On this evening, Lt Colonel Tucker, Staff Judge Advocate at Scott, was interviewed on several television newscasts, in an attempt to place the effects of sonic booms back into proper perspective. However, the late newscasts this same evening stated that officials at Scott had requested the Air Force to curtail the flights in an effort to restore better public relations between the base and the city. Two more booms occurred on this day as follows:

Time	Altitude
9.51 p.m.	43,500 feet
11.30 p.m.	44,000 feet

Friday, 6 April

Both morning and evening newspapers again gave front page attention to the booms, this time carrying statements that the booms caused the death of two rare antelopes at the zoo, which had panicked at the sound and run into the concrete wall of their cages. The papers also noted that the Air Force had been requested to curtail the flights. Once again, over seventy-five telephone complaints were received. Five more booms occurred on this day, at the following times and altitudes:

Time	Altitude
6.50 p.m.	45,000 feet
7.44 p.m.	48,000 feet
10.30 p.m.	40,000 feet
11.31 p.m.	46,000 feet
11.40 p.m.	46,000 feet

On the late television newscasts this evening both the CBS and NBC outlets announced that they had talked to the Pentagon and were informed that the B-58 flights would be stopped by May 1 because St Louis was now so familiar to the flight crews that the city was no longer of any training value. This information was later verified through official channels.

Saturday and Sunday, 7–8 April

The weekend edition of the *Globe-Democrat* carried, as its front page headline, the announcement that as of 1 May the sonic boom flights over St Louis would be discontinued.

From our experience with the sonic booms for the past ten months, those of us who have been connected with them feel that 'the exposure' was just too long. Acceptance and toleration seemed to last for a period of roughly seven months. Thereafter, in February and March, the complaints became marked by their open hostility and undoubtedly portended the events which followed in April. In our opinion, a sort of 'cumulative saturation point' was reached because the public was continuously subjected to sonic booms for such a long period of time.

References

1. A. C. Pigou, *The Economics of Welfare* (Macmillan, 1920).
2. D. L. Chadwick, Discussion of Papers in Section II, OHL, p. 167.
3. J. B. Cronin, 'A consideration of certain aspects of the law relating to noise', *Phil. Trans. Roy. Soc.*, A263 (1968), p. 235.
4. C. Wakstein and J. C. Christensen, 'Human values in engineering', *Thayer School of Engineering Newsletter* (Dartmouth College, Spring 1968).
5. K. J. Arrow, *Social Choice and Individual Values* (Yale University Press, 1963), pp. 3–4.

6. Design Research Society, 'Design participation', *Proceedings of the DRS Conference, Manchester, September 1971* (Academy Editions, London, 1972).

7. NP, Chapter 3, pp. 3–2, 3.

8. C. Wakstein, 'Hearing damage: can science be value-free?', Prototype, *Journal of Liverpool Society for Social Responsibility in Science*, Vol. 1, No. 1 (December 1970).

9. K. Saunders, Safety Officer, Safety Department, National Union of Mineworkers, London, Private Communication (Spring 1972).

10. Dr Archibald, Assistant Chief Medical Officer, National Coal Board, London, Private Communication (Spring 1972).

11. P. Doak, Institute for Sound and Vibration Research, University of Southampton, Private Communication (Summer 1967).

12. R. Hollingworth, 'American music scene news', *Melody Maker*, 22 July 1972, p. 6.

13. H. E. von Gierke and C. W. Nixon, 'Human response to sonic boom in the laboratory and the community', *JASA*, Vol. 51, No. 2 (Part 3) (1972), pp. 770, 775.

14. P. Hildrew, 'Big noises off', *The Guardian* (23 December 1971).

15. G. V. Cox, Drexel Institute of Technology, unpublished work, graduate course in Research Methods (1967).

16. L. Smedley, 'What further workmen's compensation benefits labor wants', in US Department of Labor, Bureau of Labor Standards, *Workmen's Compensation Problems, LABLAC Proceedings*, Bull. 261 (1963).

17. NP, Chapter 7, pp. 7-3, 4.

18. See for example H. A. van Loeuwen, 'Industrial deafness studies in Holland', *Phil. Trans. Roy. Soc.*, A263 (1968), pp. 271–2.

19. W. A. Rosenblith, *The Relations of Hearing Loss to Noise Exposure*, Exploratory Subcommittee Z24-X-2 of Sectional Committee on Acoustics, Vibration, and Mechanical Shock, American Standards Association (New York, 1954).

20. C. M. Harris (ed.), *Handbook of Noise Control* (McGraw-Hill, 1957).

21. C. Wakstein, Discussion of Papers in Section II, OHL, p. 168.

22. J. C. Graham, 'Practical points in the provision of hearing protectors', Brit. Acous. Soc. Meeting, University of Aston (11 October 1971).

23. G. E. Parson, Hercules, Inc., Private Communication (Summer 1967).

24. P. Kinnersley, 'Noise', *New Civil Engineer* (June 1972), pp. 46–52.

25. BBC-1, Simon Campbell-Jones, Producer (31 August 1971), 'The Noise Invasion'.

26. Pigou, op. cit.

27. A. R. Dove, HM Factory Inspectorate, Private Communication (Spring 1972).

28. NPUS, pp. 3, 7, 8.

29. Central Statistical Office, *Social Trends* (HMSO, 1970).

30. C. Wakstein, 'The Noise Problem in the United States', paper read at the Fifth International Conference on Noise Abatement, London, 13–18 May 1968; referred to as NPUS.

30a. G. Bugliarello, C. Wakstein et al., 'Noise Pollution', report sponsored by Resources for the Future, Inc., Pittsburgh, February 1968; referred to as NP.

31. Committee on the Problem of Noise, *Noise, Final Report*, Cmnd 2056 (HMSO, 1966).
32. A. Bell, 'Noise, and Occupational Hazard and Public Nuisance', Public Health Paper No. 30, World Health Organisation (Geneva, 1966).
33. American Industrial Hygiene Association (AIHA), *Industrial Noise Manual* (Detroit, AIHA, 1966).
34. Harris, op. cit.
35. A. C. McKennell, *Aircraft Noise Annoyance Around London (Heathrow) Airport*, Central Office of Information, SS. 337 (April 1963).
36. A. C. McKennell and E. A. Hunt, *Noise Annoyance in Central London*, a survey made in 1961 for the Building Research Station, SS. 332 (March 1966).
37. W. Burns, *Noise and Man* (John Murray, 1968).
38. C. Duerden, *Noise Abatement* (Butterworth, 1970).
39. K. D. Kryter, *The Effects of Noise on Man* (Academic Press, 1970).
40. Committee on Environmental Quality (CEQ), Federal Council for Science and Technology, *Noise—Sound without Value* (September 1968).
41. L. S. Goodfriend, *Noise Pollution* (CRC Press (division of Chemical Rubber Co., USA), 1972).
42. R. Taylor, *Noise* (Penguin, 1971).
43. R. Richter, *Sleep Disturbances Which We Are Not Aware Of, Caused by Traffic Noise*, EEG Station of the Neurological University Clinic, Basel (undated).
44. NPUS, pp. 3, 4.
45. International Organisation for Standardisation (ISO), 'Secretariat proposal for noise rating with respect to annoyance', ISO/TC 43 (Secretariat-272) 402 (December 1966).
46. Anon., 'Pollution by noise: how to tackle it', *Arch. Journal* (23 June 1971), p. 1444.
47. J. H. Stevens, 'Build up of the VTOL challenge', *New Scientist* (23 April 1970), pp. 172 ff.
48. J. Barr, 'The Vertiport threat', *New Society* (8 January 1970), pp. 47–51.
49. Anon., 'Anti-heliporters', *The New Yorker*, Vol. 42 (2 April 1966), pp. 37–9.
50. K. D. Kryter, 'Acceptability of aircraft noise', *Journal of Sound and Vibration*, Vol. 5 (1967), p. 364.
51. C. H. E. Warren, 'Sonic-bang studies in the United Kingdom', *JASA*, Vol. 51, No. 2 (1972) (part 3), p. 786.
52. C. B. Benjamin, Assistant Secretary, Concorde, Department of Trade and Industry, Great Britain, Private Communication (Summer 1972).
53. J. Ravetz, Department of Philosophy, University of Leeds, Private Communication (Spring 1972).
54. R. Scott-Smith, Private Communication (1972).
55. J. S. Lucas, M. E. Dobbs and K. D. Kryter, *Disturbance of Human Sleep by Subsonic Jet Aircraft Noise and Simulated Sonic Booms* (NASA CR 1780, July 1971).
56. N. Kleitman, Private Communication (Summer 1967).
57. B. H. Karplus and G. L. Bonvallet, 'A noise survey of manufacturing industry', *Am. Ind. Hyg. Assoc. Quart.*, Vol. 14 (1953), pp. 235–63.
58. *Guide for the Conservation of Hearing in Noise*, A Supplement to the Transactions of the American Academy of Ophthalmology and Otolaryngology (AAOO) (Revised 1964).

59. NP, Chapter 4, pp. 4-2, 3.
60. A. Glorig et al., *1954 Wisconsin State Fair Hearing Survey, Statistical Treatment of Clinical Audiometric Data* (AAOO, 1957).
61. ISO, 'Permissible (Low Risk) Noise Exposures for Hearing Conservation', ISO/TC 43 (Secretariat-275) 405, Draft Proposal (February 1967).
62. C. Wakstein, 'Psychological and physiological effects of noise: research and politics', paper to British Acoustical Society Meeting, University of Salford (9 December 1970).
63. W. I. Acton, Contribution to discussion of Ref. 62.
64. R. Chichester-Clark, for the Secretary of State for Employment. Written Reply to Question No. 79 by McNair-Wilson, M., House of Commons (10 May 1972).
65. Dove, op. cit.
66. L. F. Christensen, Brüel and Kjaer, Private Communication (September 1971).
67. W. P. Passchier-Vermeer, TNO, Private Communication (October 1971).
68. Department of Employment and Productivity, *Code of Practice for Reducing the Exposure of Employed Persons to Noise* (HMSO, April 1972).
69. M. Quast, Director, Openbaar Lichaam Rijnmond, Private Communication (Summer 1972).
70. Cmnd 2056, op. cit., see Ref. 31.
71. Boston Redevelopment Authority, cited as Figures 3–5 and 3–6 in Bolt, Beranek and Norman Inc., *Noise in Urban and Suburban Areas: Results of Field Studies*, Report No. 1395 to Federal Housing Administration (27 January 1967).
72. Ibid., p. 11.
73. ISO, op. cit., see Ref. 45.
74. R. H. Stearns, Thayer School of Engineering, Dartmouth College (USA), Private Communication (1968).
75. A. Alexandre, *Motor Vehicle Noise*, OECD Report U/ENV/71.9 (November 1971).
76. McKennell, op. cit.
77. M. E. Bryan, University of Salford, 'Noise Annoyance' Seminar, Mechanical Engineering Department, University of Liverpool (Autumn term, 1970).
78. Traffic Research Corporation Ltd, Toronto, *Merseyside Area Land Use Transportation Study (MALTS): The Development of a Transportation Policy Plan*, Technical Report Number 15 (May 1969), p. 188.
79. McKennell, op. cit.
80. Ibid., p. 4-4.
81. NP, Chapter 5, pp. 5-3, 4.
82. FAA, Office of Management Services, Data Systems Division, 'Enroute IFR Air Traffic Survey, Peak Day, Fiscal Year 1965'.
83. A. Alexandre, *Prévision de la gène due au bruit autour des aeroports et perspectives sur les moyens d'y remédier*, Laboratory of Allied Anthropology (Faculty of Medicine of Paris, 1970).
84. MIL Research Ltd, *Second Survey of Aircraft Noise Annoyance around London (Heathrow) Airport* (HMSO, 1971).
85. TRACOR, Austin, Texas, *Community Reaction to Airport Noise—Final Report*, document no. T-70-AU-7454-U (1970).

86. C. Bitter, *Noise Nuisance Due to Aircraft* (TNO Research Institute for Public Health Engineering, Netherlands, 1968).
87. Kryter, *Noise on Man*, op. cit., Chapter 8, pp. 239 ff, esp. pp. 293–302.
88. Ibid., 289 ff, esp. pp. 289–93.
89. MK Research, op. cit.
90. A. Alexandre, Chapter 6 in G. Bugliarello *et al.*, *Noise Pollution, a Technical, Economic and Social Problem*, Pergamon, to be published early 1973.
91. K. A. Mulholland and K. J. Attenborough, 'Predicting the noise of airports', *New Scientist* (March 1971), pp. 604 ff.
92. Commission on the Third London Airport, *The Roskill Report* (HMSO 1971).
93. TRACOR, op. cit.
94. Kryter, 'Aircraft Noise', op. cit.
95. J. C. Maxwell, Major General USAF, Director Supersonic Transport Development, Private Communication (Autumn 1967).
96. Piel, Department of Trade and Industry, Liverpool, Private Communication (Summer 1972).
97. D. Brown, ESRO, Private Communication (May 1972).
98. Wakstein, 'Noise problem in US', op. cit.
99. D. E. Hickish and J. Verbruggen, Contributions to the Discussion of paper by A. Glorig *et al.* at NPL Symposium on the Control of Noise (HMSO, 1968).
100. Committee on Hygiene Standards. British Occupational Hygiene Society, *Hygiene Standard for Wide-band Noise* (January 1971).
101. N. E. Murray, 'Hearing, impairment and compensation', *Journal of Otolaryng. Soc. Australia*, Vol. 1 (1962), p. 135.
102. N. Ashworth, Discussion of Papers in Section II, OHL, p. 167.
103. G. A. Miller and P. E. Nicoly, 'An analysis of perceptual confusions among some English consonants', *JASA*, Vol. 27 (1955), p. 338.
104. L. G. Doerfler, Department of Audiology, Eye and Ear Hospital, Pittsburgh, Private Communication (Summer 1967).
105. J. D. Hood, Discussion of Papers in Section III, OHL, pp. 238–42, 245–7.
106. R. L. Kell *et al.*, 'Social effects of hearing loss due to weaving noise', OHL, p. 187.
107. P. Ransome-Wallis, Discussion of Papers in Section III, OHL, pp. 237–8.
108. E. D. D. Dickson, 'Closing Remarks', OHL, p. 262.
109. AAOO, op. cit., see Ref. 58.
110. Murray, op. cit.
111. G. R. C. Atherley and W. G. Noble, 'Clinical picture of occupational hearing loss', OHL, pp. 193–206.
112. A. Raber, Discussion of Papers in Section II, OHL, pp. 161–3.
113. E. König, Discussion of Papers in Section II, OHL, p. 162.
114. NP, Chapter 4, p. 4-3.
115. NPUS, p. 8.
116. L. B. Lave and E. P. Seskin, 'Air pollution and human health', *Science*, (1970), pp. 723 ff.
117. B. A. Weisbrod, *Economics of Public Health* (U. Pennsylvania Press, 1961).
118. R. F. F. Dawson, *Cost of Road Accidents in Great Britain*, Road Research Laboratory Ministry of Transport, Report LR 79.

119. E. H. Blum, *Approaches to Dealing with Motor Vehicle Air Pollution*, Rand Corporation Report No. P. 3776 (December 1967).
120. H. V. Heymann, *Transport Technology and the 'Real World'*, Rand Corporation Report No. P. 2755 (June 1963).
121. D. E. Marlowe, 'Technology and society, part 1: the public interest', *Mechanical Engineering* (April 1969).
122. C. Wakstein, 'Social effects of technology', Letter to the Editor, *Mechanical Engineering* (November 1969).
123. M. E. Bryan and W. Tempest, 'Noise damage liability', OHL, pp. 143–50.
124. Design Research Society, op. cit.
125. D. L. Chadwick, Discussion on Papers in Section II, OHL, p. 169.
126. R. R. A. Coles, 'A legal action for noise deafness', *JSV*, Vol. 10 (1969), pp. 513–16.
127. C. Wakstein, 'Comments on "A legal action for noise deafness"', *JSV*, Vol. 11 (1970), p. 430.
128. R. R. A. Coles, 'Author's reply to letter by Wakstein', *JSV*, Vol. 11 (1970), 430–1.
129. D. C. Hodge and R. B. McCommons, 'Acoustical hazards of children's "toys"', *JASA*, Vol. 40, No. 4 (1966), p. 911.
130. C. Walker, 'Economics of noise', Chapter 10, NP.
131. R. Murray, Medical Adviser, TUC, Private Communication (1971).
132. F. W. Beazley, National Joint Council for the Building Industry, Private Communication (March 1972).
133. E. Marsden, Amalgamated Union of Engineering Workers, Private Communication (March 1972).
134. B. J. Weller, Civil Engineering Construction Conciliation Board, Private Communication (April 1972).
135. Anon., *Liverpool Echo* (22 August 1972).
136. G. C. Mohr, E. Guild, J. N. Cole, and H. E. von Gierke, 'Effects of low frequency and infrasonic noise on man', *Aerospace Medicine*, Vol. 36 (September 1956), pp. 817–24.
137. Z. Z. Ash'bel, 'Effects of ultrasound and high frequency noise in the blood sugar level', translated into English from Gigiena Trudi i Prof. Zabolevaniya (Moscow), Vol. 2 (1956), pp. 29–32, JPRS—36252.
138. W. Tempest and M. E. Bryan, 'Low frequency sound measurement in vehicles', *Applied Acoustics*, Vol. 5 (1972), p. 139.
139. J. Setaloff, *Hearing Loss* (Lippincott, Philadelphia, 1966), Chapter 25.
140. D. A. Ramsdell, 'The psychology of the hard-of-hearing and the deafened adult', Chapter 18 in H. Davis and S. R. Silverman, *Hearing and Deafness* (New York, Holt, Rinehart and Winston, 1961).
141. G. Jansen, 'Zur Entstehung Vegitativer Funktionstorungen durch Larminwirkug', *Archiv. Gewerbepath. u. Gewerbehyg*, Vol. 17 (1959), pp. 238–61.
142. J. P. Connell, The Noise Abatement Society, 'The biological effects of noise', paper given at BASS Annual Meeting (5 September 1972), pp. 3–4.
143. Cmnd. 2356, op. cit.
144. I. Abey-Wickrama *et al.*, 'Mental hospital admissions and aircraft noise', *The Lancet* (13 December 1969).
145. See Ref. 85.
146. D. Broadbent, NC, Chapter 10, pp. 10–18.

147. O. C. Herfindahl and A. V. Kneese, 'Quality of the environment', *Resources for the Future*, Inc. (1965), p. 17.
148. J. Ravetz, 'Science and values', a series of radio discussions, Radio 3 (Winter 1972).
149. R. Sherrill, 'The jet noise is getting awful', *New York Times Magazine* (January 14 1968).
150. Office of Science and Technology (OST), Executive Office of the President, *Alleviation of Jet Aircraft Noise Near Airports, A Report of the Jet Aircraft Noise Panel* (March 1966).
151. McKennell, op. cit.
152. Ibid.
153. K. D. Kryter and K. S. Pearsons, 'Laboratory tests of subjective reactions to sonic booms', *NASA* CH-187 (March 1965).
154. D. E. Broadbent and D. W. Robinson, 'Subjective measurement of the relative annoyance of simulated sonic booms and aircraft noise', *JSV*, Vol. 1 (1964), p. 162.
155. C. W. Nixon and H. H. Hubbard, 'Results of USAF-USA-FAA programme to study community responses to sonic booms in the greater St Louis area', *NASA* TN D-2705 (May 1965).
156. P. N. Borsky, *Community Reactions to Sonic Booms, Oklahoma City Area, February–July 1964, Part 1. Major Findings*, National Opinion Research Center, University of Chicago, 55 Fifth Ave, New York, Report No. 101 (January 1965).
157. *Sonic Boom Experiments at Edwards Air Force Base, Interim Report*, National Sonic Boom Evaluation Office, 1400 Wilson Boulevard, Arlington, Va., NSBEO-1-67 (28 July 1967).
158. Kryter and Pearsons, op. cit.
159. C. Wakstein, 'Preventive technology', to be submitted to *New Civil Engineer*.
160. T. C. Sinclair, Science Policy Research Unit, University of Sussex, Private Communication (1971).
161. J. Cumming, Department of Mechanical Engineering, University of Liverpool, unpublished research for the degree M.Eng., 1970–1.
162. Nixon and Hubbard, op. cit., p. 551.
163. Ibid.
164. Borsky, op. cit.
165. Kryter, 'Aircraft noise', op. cit.
166. D. A. Hilton *et al.*, 'Sonic boom exposures during FAA community response studies over a 6-month period in the Oklahoma City area', *NASA* TN D-2539 (December 1964).
167. A. E. Knowlton (ed.), *Standard Handbook for Electrical Engineers* (McGraw-Hill, 1949), Fig. 11-4.
168 Lucas *et al.*, op. cit.
169. C. H. E. Warren, 'Experiments in the United Kingdom on the effects of sonic bangs', *JASA*, Vol. 39 (1966), p. S59.
170. C. Wakstein and S. J. Kennett, *Engineering Failures* (Longmans, to be published), Chapter 2.
171. See Ref. 157.
172. A. H. Odell, 'Jet Noise at John F. Kennedy International Airport', in Ref. 150.
173. J. Barnes, Chapter 27 in Bugliarello, op. cit.
174. Odell, op. cit.

175. British Aircraft Corporation, 'Concorde supersonic flight test routes and the sonic boom', c. 1969 (undated, CONFIDENTIAL), pp. 18, 20.
176. Crockett, cited in C. Wakstein, 'Review of the Fifth International Conference for Noise Abatement, London', *JASA*, Vol. 15 (1969), pp. 256–7.
177. McKennell, op. cit.
178. Crockett, Private Communication (1968).
179. B. L. Clarkson and W. H. Mayes, 'Sonic-boom induced building structure responses including damage', *JASA*, Vol. 51, No. 2 (part 3) (1972), p. 754.
180. Institute for Defense Analyses, *Demand Analysis for Air Travel by SST*, AD 652309, 10, Table J-2.
181. J. C. Maxwell, Major General USAF, Director Supersonic Transport Development, Private Communication (Summer 1967).
182. Kryter, 'Aircraft noise', op. cit.
183. Ibid.
184. Nixon and Hubbard, op. cit.
185. Clarkson and Mayes, op. cit.
186. B. K. O. Lundberg, 'The acceptable nominal sonic boom overpressure in SST operation', paper presented at the National Conference on Noise as a Public Health Hazard, Washington D.C. (June 13–14 1968), Fig. 1.
187. Clarkson and Mayes, op. cit.
188. Kryter, *Noise on Man*, op. cit., p. 455.
189. *Demand Analysis*, op. cit.
190. Federal Aviation Administration, *Economic feasibility report: United States supersonic transport* (1967) (AD 652 313).
191. H. Raiffa, *Decision Analysis* (Addison-Wesley, 1968).
192. J. Ardill, 'Call to end internal passenger flights', *The Guardian* (23 March 1972).
193. 5th International Conference on Noise Abatement, London, Private Communication (May 1968).
194. D. E. Wooster, Transportation Research Institute, Carnegie-Mellon University, Private Communication (Summer 1967).
195. ISO, op. cit., see Ref. 45.
196. See Ref. 78.
197. US Census of Housing 1960.
198. See Ref. 78.
199. Automobile Manufacturers' Association, *Motor Truck Facts*, 1967 Edition, p. 16.
200. R. A. Wason, 'Quiet replacement truck mufflers', *Noise Control*, Vol. 2 (May 1956), p. 44.
201. Ibid.
202. Alexandre, op. cit., p. 70.
203. Bugliarello, op. cit., Chapter 9, Fig. 18a.
204. Alexandre, op. cit., p. 61.
205. Ibid., p. 28.
206. Parson, op. cit.
207. Ibid.
208. CEQ, op. cit.
209. *The Iron Age* (Metalworking Annual, 1963).
210. US Department of Commerce, Bureau of the Census, Census of Manufacturers 1963.

211. T. Bonney, Director of Industrial Hygiene, Aluminium Corporation of America, Pittsburgh, Private Communication (1967).
212. M. E. Delany, 'Some sources of variance in the determination of hearing level', OHL, pp. 97–108.
213. J. L. Belser, Continental Can Company Inc., Private Communication (Winter 1968).
214. C. Wakstein, Discussion of Papers in Section II, OHL, p. 168.
215. J. C. Kell et al., 'Social effects of hearing loss due to weaving noise', OHL, p. 174.
216. A. N. Wakstein, Birmingham School of Speech Therapy, Private Communication (Summer 1972).
217. F. A. van Atta and C. Wakstein, Discussion of Papers in Section II, OHL, pp. 164–5.
218. van Atta, op. cit. 'It is one thing to have a law and another to have some money for the enforcement of that law. Under the Walsh-Healey Act, which is the one we use most, and the Service Contract Act, we do not regulate as a matter of law but as a matter of contract. The only sanction which we have is to cancel the present contract or to deny future contracts. We have very little use for enforcement under these laws. Our manpower is quite small. We go into about 3,000 plants a year and out of those 3,000 inspections we normally have about three legal actions which go to a hearing. One reason that we have only three is because we never lose a case: the penalty is always the same, denial of Federal contracts for three years. For a company in the supply business this is an unhappy thing.'
219. A. R. Dove, 'Hearing protection in the factory invironment', Brit. Acous. Soc. Meeting, University of Aston (11 October 1971).
220. P. Gillman and A. Woolf, 'The dangerous dust', The Sunday Times Magazine (2 April 1972), p. 34.
221. Ibid., note also the reluctance of the Factory Inspectorate to apply to a magistrate to close the factory in question.
222. Kryter, Noise on Man, op. cit., pp. 436–9.
223. Ibid., pp. 439–41.
224. See Ref. 78.
225. Kryter, Noise on Man, op. cit., pp. 442–3.
226. See Ref. 68.
227. Quast, op. cit.
228. Heymann, op. cit.
229. Blum, op. cit.
230. C. W. Churchman, Challenge to Reason (McGraw-Hill, 1968).
231. C. Wakstein, 'The Liverpool M62 public participation exercise: the public's limited view of its own rights', to be submitted to New Civil Engineer.
232. R. L. Ackoff, Scientific Method: Optimising Applied Research Decisions (Wiley, 1962), p. 82.
233. E. W. Davis, 'Comparison of Noise and Vibration Levels in Rapid Transit Systems', Report prepared by Operations Research Inc. for National Capital Transportation Agency.
234. Anon., New York Magazine (Summer 1972).
235. Atherley and Noble, op. cit.
236. D. E. Broadbent, Medical Research Council Applied Psychology Unit, Private Communication (January 1972).

237. K. D. Kryter *et al.*, 'Hazardous exposure to intermittent and steady-state noise', *JASA*, Vol. 39 (1966), pp. 451–64.
238. R. K. Ham, 'The electricity supply industry' in *The Man Made World* (The Open University Press, 1971), Fig. 1.
239. J. O. Cookson, Machine Tool Industry Research Association, Private Communication (January 1972).
240. Karplus and Bonvallet, op. cit.
241. C. N. Howison, Air Pollution Control League of Greater Cincinnati, Private Communication (Summer 1967).
242. W. Burns and D. W. Robinson, *Hearing and Noise in Industry* (HMSO, 1970).
243. W. L. Baughn, 'Noise control—percent of population protected', *Int. Audio.*, Vol. 5 (1966), pp. 331–8.
244. J. Connell, 'Noise Torch', letter to the editor, *The Guardian*, Manchester (18 May 1972).
245. See Ref. 236.
246. M. Meyerstein, University of Liverpool Mechanical Engineering Department, Final Year Project Report, 1971.
247. P. N. Borsky, National Opinion Research Center, 'Community reactions to sonic booms', Report No. 87 (August 1962), CONFIDENTIAL.
248. The Committee on Medical Rating of Physical Impairment, 'Guide to evaluation of permanent impairment; ear, nose and throat and related structures', *JAMA*, Vol. 177, No. 98 (August 1961), pp. 489–98.
249. M. S. Fox, 'Compensatory provision for occupational hearing loss', *Arch. Otolaryng.*, Vol. 81 (March 1965), p. 257.

Key:
OHL = D. W. Robinson (ed.), *Occupational Hearing Loss* (Academic, New York, 1971).
JASA = Journal of the Acoustical Society of America.
JSV = Journal of Sound and Vibration.

Chapter 11

The Law Relating to the Regulation of Noise

by WILLIAM C. OSBORN
Public Interest Research Centre Ltd

The most remarkable feature of legal sanctions to control noise is that they generally have not worked. Noise levels are increasing around the world, and ironically the rise is most rapid in those countries considered to have the highest standard of living. In the United States the over-all loudness of environmental noise is doubling every ten years [1]. In New York City, one of the noisiest places on earth, it is increasing by a decibel a year.

In most countries, the law has failed to prevent noise at its source, to abate it once it has occurred, or even to compensate adequately its proven victims. Noise is on the increase because the law has not put a price on noisemaking commensurate with its true social cost. Until noisemaking becomes expensive, the potential of technology for quietening the machine will go unused and the general public will continue to bear the costs in declining health and amenity.

In this chapter I shall sketch briefly the existing types of legal remedies available to combat noise pollution and describe why I think they have been unsuccessful. I shall then make some proposals for changes which I consider essential to the efficacy of any noise control programme. The chapter concentrates chiefly on those areas of British and American law which have equivalents in other countries. Little is said about international controls for, except in isolated cases, they do not yet exist.

Existing Laws and Their Failures

Noise control laws can be divided into three categories: (i) private right of action; (ii) local or central anti-noise ordinance; and (iii) miscellaneous preventive provisions such as zoning, building codes and noise emission standards.

The weakness of the measures in all these categories derives essentially from two related problems. First is the nature of the sanctions themselves. Except for some of those in the last group, most laws are retrospective in concept; instead of seeking to prevent noise before emission, these sanctions attempt to abate noise after it has occurred. This is an inefficient approach to the problem. Correcting an existing faulty technology is more expensive and troublesome than designing initially for quiet operation.

The second problem concerns enforcement. Even this first shortcoming could be overcome if vigorous enforcement were undertaken. The laws would then deter the making and operating of noisy machinery. Like other environmental laws however, noise control measures have been woefully underenforced. Because fines are low, and enforcement mild and piecemeal, it pays the 'polluters'—the producer and the user of noisemaking machinery —to continue the din.

Private Right of Action against the Noisemaker

In common-law countries the oldest remedy against noise is the private suit for nuisance. Countries without a common-law tradition usually permit such individual action by statute.

In these suits the victim of excessive noise asks a court for damages to cover the noise-induced harm or for an injunction against the noisemaker, or both. Where the plaintiff can show substantial interference with his person or with the enjoyment of his property, he may be able to recover [2]. Effective in some cases, this remedy usually founders on obstacles which prevent it from contributing significantly to noise abatement efforts.

In defining nuisance, courts will usually balance what they believe to be the social utility of the activity which produces the noise against the harm done to the plaintiff. Thus, even if the plaintiff can prove injury, a court often decides that the social value of the noisemaking enterprise is of greater importance and denies relief. The problem frequently arises in the case of local industrial works which offend many people in addition to the plaintiff, but contribute to the area's economy. Such balancing amounts to judicial capitulation to noisy technology and an affirmation of the notion that noise is the to-be-endured price of progress.

The plaintiff may prevail if he can show that the noise interferes with him in a different or more offensive manner than it does with anyone else. However, such harm is judged by objective rather than subjective standards. A plaintiff with special sensitivity to noise cannot normally win unless he can show that the average man would be offended by the same noise.

Embarrassed by making life so hard for him, the law in some jurisdictions gives the plaintiff who surmounts these obstacles a sort of bonus: it is no defence to show that all reasonable steps have been taken to prevent the noise [3]. This important protection for plaintiffs has unfortunately been discarded in newer statutes.

Another difficulty with a private action is that it can really only be effective against a single stationary source. It is virtually useless against many of the types of noise which are most offensive today, such as noise from moving sources, noise of short and intermittent duration and general noise from several sources. The plaintiff must carry the burden of proving that the noise he complains about did or does come from the defendant. This burden is onerous enough in a case with a single stationary source. It becomes nearly impossible where there are many potential sources.

A further problem with the private nuisance action is that an individual or group of individuals must initiate and pursue it. Added to the plaintiff's burden of proof and the balancing done by the court, a nuisance action is long and expensive, especially where *champerty** is forbidden. These facts do not encourage the average noise-aggrieved citizen to file suit.

Dependence on individual initiative is at the same time the action's great strength, for when all other abatement laws have failed, as they often do, common-law rights persist and can theoretically be taken up by any injured party. It is a sad and ironic testimony to the potential strength of the common law in Britain that Section 41(2) of the Civil Aviation Act 1949 prohibits civil action against aircraft noise at airports.

The ultimate weakness of the common-law nuisance action is that it must be initiated after the noisemaking has begun, usually after considerable investment in a technology which has never been designed for quiet. Consequently, it only indirectly encourages innovations and advances towards quieter machinery. Although occasionally a useful remedy, it treats only the symptoms of the noise disease, and does not begin to prescribe a programme of prevention which other regulatory systems can give.

Anti-noise Ordinances

The great bulk of noise-control law around the world takes the form of ordinances. These laws typically prohibit excessive noise-

---

* Illegal proceeding whereby one party agrees to assist a litigant on condition that if the suit is successful the disputed property will be shared between them.

making, provide for the apprehension and prosecution of violators, and prescribe penalties. They occur at almost every governmental level. In the United States, most municipal jurisdictions,
and many state governmental bodies have passed noise control
laws of this type. European countries have similar laws, and in
Britain, a national law, the Noise Abatement Act of 1960, makes
excessive noise a statutory nuisance subject to abatement by local
authorities and magistrates' courts.

Anti-noise ordinances define excessive noise in a number of
ways. The older laws forbid 'unnecessary or unreasonable' noise.
The British law, and some like it in America, borrow the definition
for statutory nuisance from English common law. Some recent
laws set decibel limits on noise-making activities.

The vagueness of the first type of statute benefits the noisemaker, particularly large industries. For what might seem unreasonable or unnecessary to a victim of the noise might not seem
so to an administrative official or judge. As in deciding private
nuisance actions, courts typically use a balancing test in inter-
preting these ordinances, and, in weighing the social usefulness of
the noisemaking activity, tend to favour entrenched economic
interests.

Many anti-noise laws go a step further than the common law
in coddling the noisemaker. For instance the British Noise Abatement Act offers a new defence to the person charged; if he is
found to be applying the best practicable means of controlling
the noise, then he will not be prosecuted even though his noisemaking would otherwise constitute a nuisance. This defence, which
further dilutes the rights of the victims of excessive noise, has been
expressly rejected at common law, but appears in a number of
British statutories, such as the Public Health Act of 1936 and the
Alkali Acts.

The 'best practicable means' are defined as what is technically
possible having regard to costs of control and to local conditions
and circumstances. This sounds reasonable enough and seems to
grant a sensible defence to the noisemaker while still assuring the
public that everything within the realm of the possible will be done
to protect them. As the guardians of the common law probably
understood, it is in practice that the 'best practicable means' idea
collapses however and undermines the aim of the law. For this
phrase usually means what the noisemaker says he can afford,
and what he says he can afford is not a matter which most government officials are particularly well-equipped to evaluate. Besides
this, the concept contains another, perhaps even more serious

fault being basically a passive mechanism. Instead of placing the onus on the noisemaker to discover and develop new control methods, the burden of enduring the noise is on the public while the polluter waits for someone else to come up with the answer.

The regressive notion embodied in the 'best practicable means' idea is not unique to Britain. Equivalent concessions to faulty technology and the vagaries of corporate accountancy exist in many national laws. In most US codes, the noisemaker is not forced to install abatement equipment if the installation and maintenance would be technically or economically unreasonable. And even in the Soviet Union, where sophisticated noise standards existed in law long before their appearance in the West, the factory manager can plead inadequate technology or finances as a valid defence to government officials from other departments charged with abating the noise from his factories.

Laws which set decibel limits relieve enforcement officials of the task of defining excessive noise. Too often, however, the limits which are set merely tend to legitimise existing levels and put no pressure on noisemakers to become quieter.

Many anti-noise laws have been further emasculated by provisions which put major noise sources beyond their control. Thus the Noise Abatement Act in Britain does not apply to aircraft, motor vehicles or statutory undertakers (which include British Rail, the British Airports Authority and the nationalised iron and steel industry). In the US, the attempts of towns adjacent to airports to gain relief from aircraft noise by passing ordinances prohibiting overflights have been nullified in court on the grounds that these conflicted directly with the Federal law regulating aircraft flight [4]. In effect, then, local victims of aircraft noise in the US, as in Britain, have been forced to look for relief to their national government. As will be seen though, the US government, timid about offending the mighty airlines, has done precious little to protect the public.

Inadequate enforcement of anti-noise laws is one of the major reasons for their ineffectiveness. Not only are the laws difficult to administer, but responsibility for their enforcement is usually delegated to an understaffed and overworked police or public health department, which often has neither the time, experience nor inclination to find and prosecute noisemakers. On top of this, low penalties found in most anti-noise law do little to encourage general compliance.

Like the common-law nuisance action, the average anti-noise law is difficult to apply because so much modern noise is a mixture from several sources or results from transient or mobile activity. Generally, the more specific the ordinance, the more these problems are overcome and the simpler enforcement becomes. Thus, laws against horn-blowing have been successful in several cities. Even laws which set specific decibel limits, however, such as vehicle operation noise laws, present to enforcement officials the challenge of identifying violators and isolating them from many other noise sources.

A recent American case offers new hope to officials stymied in their efforts to enforce noise laws against multiple sources. In *State* v. *Dorset* (1968) the defendant motorcyclists were convicted of violating a municipal ordinance which proscribed unreasonably loud noises. Each defendant argued that he separately did not violate the ordinance. The court held, however, that where the group noise violated the statute, individuals who made that noise were responsible for it even where it was not established how much each one contributed. The decision in effect shifts the burden of proving causation from the state and could be a step towards general apportionment of liability among several suspects none of whom can be established as directly responsible. A proposed abatement act for New York City adopts this concept by giving government authority to shut down or order abatement of a device which may not itself violate the law but which contributes to a cumulatively harmful set of noise sources. If other municipalities and common-law courts embraced this theory, enforcement would be greatly simplified.

Most anti-noise ordinances depend for enforcement on government officials who cannot give noise control the attention it needs and who are vulnerable to pressure from powerful noisemakers to be lenient. Except in a few cases, the laws do not attempt to ease this burden on government by permitting enforcement actions by ordinary members of the public. Statutory nuisance laws normally do allow citizen action, but often create pointless obstacles to its effective use. For instance, Britain's Noise Abatement Act requires that three noise-aggrieved citizens must sign a complaint before a magistrate will hear it.

The penalties available under most anti-noise law are so minimal that, even overcoming the enforcement problem, they could hardly be expected to deter violators. Most statutes treat violations as misdemeanours and set maximum fines so small in most cases that a large transgressor could easily afford to pay rather than install quietening devices. For instance, the maximum penalties in the Noise Abatement Act are £20 for failure to comply with an abatement or prohibition notice and £50 for failure to comply with a nuisance order with £5 per day additional for a continuing violation. Injunctions are rarer than fines and few laws, if any, provide for imprisonment against a violator.

Many of the shortcomings of the laws reflect their irremedial weakness in dealing effectively with the noise crisis, they approach noise pollution *ex post facto*. Although important ingredients in any noise control effort, they should not be relied on to the exclusion of more preventive measures such as source emission standards.

Noise in the Workplace

While on the subject of ordinances, a word should be said about a special class of environments to which these types of law sometimes apply.

Industrial workplaces offer opportunities for noise control by ordinances which are absent in the general environment—a limited, measurable space; easily ascertained noise sources; and legal control of the environment in the hands of a few known individuals. Industry, too, either has, or has access to, enormous technical resources for solving the problem presented by excessive noise in factories. Considering such impressive advantages, one might expect that noise control schemes in industrial environments would have reached a degree of success unparalleled by those in the general environment. Unfortunately, this has not been the case. Noise in industry is a pandemic problem and shamefully little is being done to quieten things.

The World Health Organisation recently estimated [5] that the annual monetary loss due to accidents, absenteeism, inefficiency, and compensation claims attributable to industrial noise in the United States is 4 billion dollars. The number of American workers experiencing noise conditions unsafe to hearing is estimated to be in excess of 6 million and may be as high as 16 million [6]. In the UK things are not much better. Noise injury research has concluded that human exposure to 85 dB(A) for eight hours a day can cause hearing damage. In his 1969 Annual Report, the Chief Inspector of Factories published the results of noise level monitoring carried out by the Inspectorate in several industries. Although little was said about the period of exposure, an alarming number of measurements exceeded 90 dB(A) and some went above 110 dB(A).

In spite of such statistics, industry and government in most countries have moved with glacial slowness to abate factory noise. Even the trade unions have accorded the problem a low priority.

The response of the average industrialist or works manager to a noise problem is to order ear muffs or plugs for the workers and forget about it. If the worker suffers, then it's his own fault. This primitive attitude ignores the fact that intense noise or vibration can affect more than simply a man's hearing. It also ignores the fact that often ear protection is impracticable: men must sometimes talk to one another, and must be able to hear safety alarms; ear plugs are uncomfortable if not fitted properly; and ear muffs, unbearable in the heat frequently found around noisy machinery. Finally, this approach avoids altogether the challenge of making factory machinery quieter, often when the technology for this exists.

Some managers recognise these problems and rotate men through the noisy jobs in the works. While this policy affords some immediate relief to the worker, it avoids the more desirable engineering solution and still risks the unknown dangers of short-term exposure to intense noise.

The trade unions' record in fighting industrial noise has been little better than that of management. Action is often limited to a demand for more money for workers who have to toil in a noisy place. While a man certainly should receive a higher wage for hazardous work, this is no solution to factory noise. If given the choice, most people would elect not to work in a noisy environment if it meant hearing loss, no matter how high the wage. Yet this choice is never offered to workers. Hazardous duty pay is accepted as the only answer. The increased wage will rarely, if ever, cover the actual value of hearing loss. Furthermore, unless it is much higher and more widely applied than it tends to be in fact, it will not induce management to make things quieter in order to avoid the extra payroll cost.

Fortunately, some trade unions are now taking a stronger stand against industrial noise. The recent backing by the General and Municipal Workers' Union of a power tool operator who successfully sued his employer for deafness caused by noise at his work, is a hopeful sign [7].

The failure of government to protect the working man from industrial noise is scandalous. In some of the most highly industrialised countries no enforceable standards for the exposure of workers to noise exist at all. Despite the figures in the UK Factory Inspectorate's 1969 report, no standard for British workplaces has yet been set.

The Americans took a novel, if limited approach to the problem, but successive administrations ended up betraying the law's intent. Under an old law, the Walsh–Healey Public Contracts Act [8], the Secretary of the Department of Labour was authorised to set standards for industrial health and safety in the workplaces of firms doing more than $10,000 worth of business annually with the US government. Although this power existed as long ago as 1942, the Secretary did not promulgate specific noise standards until 1969. Even then, the limits were set *above* the level at which research had shown hearing harm to occur! In a last minute concession to industry, the legal limit was set at 90 dB(A).

Through new Federal legislation, the 1970 Occupational Safety and Health Act [9], the government now has the authority to regulate noise in most industrial environments. At least for the time being, the new Act has adopted the old Walsh–Healey standards. The basic weaknesses remain: the standard limits ambient noise and does not set emission limits for factory machinery; it contains nothing in the way of a margin of safety, and in fact permits noise exposure proven to be harmful; finally, it does not provide for greater quiet in the future through graduated standards. However, regulations published recently under the Act suggest that the government will begin to look with disfavour upon the old management attitudes [10]. A manager must now use 'feasible administrative or engineering controls' to reduce noise. If these controls fail, he may then fall back on personal protective equipment.

In contrast to Britain and the United States, the Soviet Union seems, at least on paper, to have taken an earlier and more comprehensive interest in workplace noise. Since 1956, the Ministry of Health has promulgated Sanitary Norms for industrial noise, and these also cover industrial noise emissions to adjacent areas. Although the maximum permissible exposure amounts to but slight improvement on the American standard, the norms do impose stricter limits for higher frequency noise and different standards for different types of workplaces. Furthermore, under a new scheme, the Russians require labelling of the noise characteristics of certain industrial machinery and have even started to set emission standards for some equipment [11].

Enforcement of occupational noise standards has been minimal. In the US under the Walsh–Healey Act, failure to abide by the Labour Department's standards drew a maximum penalty of loss of the right to contract with the Federal government for three years. Although perhaps a significant setback for a firm having a major fraction of its business with the government, such a sanction was rarely imposed. The reason was not that everyone was complying with the law, but rather that there was virtually no monitoring or enforcement done by the government. The Department of Labour had less than forty inspectors to cover over 27,000,000 employees in 75,000 workplaces, thus many hazardous spots were never visited. Even when an inspector answered a complaint he would inform the works of his visit in advance. The report he wrote and submitted together with the findings from his inspection would invariably be withheld from the workers' representatives, so that no check could be made on what the inspector reported.

Enforcement in the new Occupational Safety and Health Act is not completely resolved. However, the law forbids advance warnings of inspections to firms and, it is hoped, will be administered in a more open manner than was its predecessor. The Act carries stiffer penalties—up to $10,000 for a violation of a standard—but there is no guarantee that the deterrent potential of these will be realised.

If industrial health and safety laws have failed to protect workers from excessive noise, compensation schemes for those who have lost their hearing have provided even less.

In many countries, occupational hearing-loss has only recently been recognised by the law as a compensatable industrial injury. Even in those areas where deafness is compensatable, the benefits and how they are awarded offer little relief to the worker and minimal incentive to the firm to make things quieter.

In America, even though the potential cost of compensation for industrial hearing loss is alarmingly high, the actual number of claims paid is small. One estimate is that if only 10 per cent of those workers eligible filed a claim and won an average award of $1,000, a total of $450 million would be paid [12]. In fact, one source has estimated that fewer than 500 claims per year are paid for occupational noise-induced hearing loss. The 1968 report of the Committee on Environmental Quality lists some reasons for the small number of claims: 'Many afflicted workers do not know that their hearing loss is compensatable. Compensation laws in some states honour claims for total deafness but not the more usual partial loss of hearing. Workman's compensation provisions in other states cover partial loss of hearing but require the claimant to be six months away from the job before settlement can take place' [13]. In addition, because compensation claims that are won increase the payments employers must pay into the compensation fund, most firms will devote considerable energy fighting each compensation claim filed. Many large firms employ experienced lawyers who specialise in compensation law and spend all their time fighting claims. The average worker seldom has such resources, and as the law puts the burden of proving the claim on him, can rarely succeed.

Even where a claim prevails, the benefits usually offered by the law are meagre by any standard. In New York, for instance, the maximum benefit for hearing loss is $80 per week for sixty weeks for one ear, 150 weeks for two (1969 figures) [14].

Until the road is made easier by removal of these obstacles for the worker seeking compensation and more expensive for the employer running a noisy works by creating incentives for quieter workplaces, the social costs will continue to be misplaced.

Zoning, Building Codes and Emission Standards

Zoning laws, building codes and emission standards avoid the inefficient, retrospective aspect of both the anti-noise ordinance and the private damage suit and offer the greatest hope in the search for effective regulation. Unfortunately however laws in this third category have not lived up to their potential.

Zoning and planning laws assume a number of forms. In the United States, most local governments have statutory authority empowering them to restrict various activities in their areas. Liberal variance provisions, however, usually make zoning procedures vulnerable to the political power of large industry. Furthermore noise control has never been a high priority in the administration of zoning laws. A recent report of the US Environmental Protection Agency to the President and the Congress on Noise concluded that:

'Inclusion of noise standards in zoning codes is generally recent, and few are well enforced. Many cities with quantitative noise limits in zoning codes have no measuring equipment for enforcement purposes and there is again a need for guidelines in formulating workable standards. Standards are useful for planning and zoning commissions in screening applicants for industrial locations' [15].

Control through zoning and planning has fared little better in the UK. Under their development powers local authorities can refuse planning permission to an industrial development they believe will create a noise problem. The apparent failure of some local authorities to use this power responsibly has prompted the

Noise Advisory Council (NAC) to recommend that 'it should be a duty of local authorities in the exercise of development control powers to take account of the noise implications of proposed new development (among other factors) in arriving at their decisions' [16]. Despite this good sense, however, the working group later rejects a suggestion that industrialists be required to notify local authorities of proposed installations of equipment or changes in working methods likely to be noisy because the benefits 'would be outweighed by administrative complexities'. If we are to encourage preventive laws and procedures, we must not be put off by such things as 'administrative complexities'. In this case notification prior to installation, if accompanied by a provision authorising local authorities to attach conditions, as they could with a planning permission application, would be a good way of nipping new noise problems in the bud.

The NAC, reflecting the views of the Association of Public Health Inspectors, recommends that local authorities be given the power to designate Noise Abatement Zones, and that, subject to confirmation by the Secretary of State for the Environment, they be permitted to specify the noise level for the zone in which they could restrict any noise emissions. These are good ideas and the NAC legislative proposals containing them should be supported, provided they contain the types of procedural and enforcement safeguards discussed below.

In the Soviet Union two types of zoning standards exist to control noise from industrial facilities. The Sanitary Norm of 1969 for industrial noise, already mentioned, covers noise levels in industry and industrial noise emissions to adjacent communities. The other is the sanitary protective zone which limits noise levels emanating from industries in certain designated areas. The effectiveness of both schemes has been vitiated by loopholes and enforcement problems. Like their Western counterparts, Soviet factory managers have little incentive to operate quietly and when confronted with a violation can request an exemption on the grounds of inadequate development of technology. Enforcement of noise standards as elsewhere appears to be of low priority in the Soviet Union, with enforcement officials overworked and responsible for many problems.

In the United States the vast power of the Federal government establishment is being used more frequently to achieve compliance with environmental goals. At least three Federal laws contain requirements which could lead to noise suppression at the planning stage of major public works projects. The most general is the

National Environmental Policy Act of 1969 [17] which attempts to build an active awareness of environmental problems into governmental decision-making and requires departments and agencies of the Federal government to produce environmental impact statements on all Federal actions significantly affecting the human environment before these actions commence. Although noise has not yet become a major issue in the controversies over impact statements, several government projects have been successfully halted in court by environmental groups who have claimed that impact statements did not adequately lay out the problems or the solutions raised by the proposed government action.

Two major types of federally aided projects are highways and airports, and proposals for both require environmental impact statements. In addition, the Federal-Aid Highway Act [18] and the Airport and Airways Development Act place restrictions on governmental approval of these projects. The highway legislation authorises the Secretary of Transportation to withhold approval of highways until specifications include adequate implementation of appropriate noise standards. The second law requires the consideration of the environmental impact of airport and aircraft activity on adjacent communities and provides for public hearings on airport projects.

While an important step towards bringing noise control into planning decisions and affording some relief, these laws cannot achieve very much by themselves, unless strict emission and operating standards are at the same time reducing the source noise of vehicles and aircraft. Ideally, the laws should work to encourage stricter noise emission standards. For example, if the highway noise standards under the Federal-Aid Highway Act were so strict that compliance could only be achieved through quieter vehicles, then the withholding of Federal approval and support for highways would form a powerful incentive to develop quieter vehicles. Given the normal timidity of enforcement officials, however, the laws are unlikely to be used in this way.

Building codes are another type of planning law, but the emphasis again is not on suppressing noise at the source but rather in lessening its impact. Building codes typically set interior levels for residential and business premises which architects and construction engineers must achieve through proper location and use of building materials which can also be specified directly.

Most European countries have such codes and in Britain a British Standards Institution sets out in some detail specifications for buildings which relate to noise transmission [19]. Perhaps the most comprehensive building codes are to be found in the Soviet Union. A sanitary norm for interior noise levels in housing has existed since before 1960. The latest regulation, SN 535-65, specifies limits both inside buildings and outside in the communal land of the apartment complex. The basic norms are adjustable, depending on the location of the housing development—suburban, urban, or within sanitary protective zone; proximity of traffic; quality and duration of the noise; and time of the day the noise occurs. For each of these variables, one adds or subtracts a specified number of decibels to arrive at the standard appropriate to the special circumstances of the particular housing unit.

One shortcoming is the failure to protect people living near main city or inter-city highways. Instead of requiring added insulation to preserve the interior environment from the increased outside noise, the particular adjustment allows *more* noise inside by raising the norm for units facing the road.

This policy may be based on economics or on an erroneous belief that the sound volume inside could be turned up without consequence when the volume outside increases. In any case, such a slippage destroys the practical impact of the regulation on some of the noisiest living and working quarters.

Coupled with these interior noise standards are design and construction standards which set out detailed acoustical specifications, such as soundproofing and the location of utilities.

Similar regulations exist in the US, though not nearly as comprehensively as in the Soviet Union. The Federal Housing Administration sets minimum property standards for dwellings it helps to finance [20]. The standards wisely permit state or local agencies to set even stricter requirements. Until recently though, the FHA, in a discriminatory policy as shortsighted as that just described, permitted noisier interiors in urban buildings surrounded by high exterior noise levels than in quieter suburban settings.

Acoustical standards for buildings could contribute significantly to quieter indoors for everyone, if they (i) are stringent enough to encourage the development of new and quieter building technologies, rather than simply to perpetuate existing levels; (ii) discard the faulty logic of tailoring interior noise standards to background exterior noise levels; and (iii) are accompanied by monitoring and enforcement measures ensuring compliance.

Noise Emission Standards

Presently limited in use, noise standards for new machinery offer great hope of preventing noise at the source. Typically, laws containing standards set decibel limits on the noise a product is allowed to emit before it can be sold or operated. California, for instance, limits noise emissions of motor vehicles to be sold in the state [21]. Other states and even some municipalities are following California's lead and are even expanding emission standards to cover other products. There will soon be a Federal Noise Control Act embodying the emissions approach and authorising the Environmental Protection Agency to set standards for certain products [22]. Whether or not the Federal law, when it is passed, will supersede and pre-empt the noise standards of other jurisdictions is a question of considerable controversy which will be discussed below under the section on local versus central control.

In the United Kingdom, under the Motor Vehicles (Construction and Use) Regulations of 1969, the Secretary of State for the Environment has promulgated standards for new vehicles. Also, the Working Group on the Noise Abatement Act of the Noise Advisory Council, in its recent report on neighbourhood noise, suggests that the government impose statutory requirements on manufacturers.

Noise law in the Soviet Union, too, is developing along the lines of emission standards, with the USSR Committee on Standards considering maximum emission standards for hand tools, metal-cutting machine tools and electric motors.

One great problem with existing emission control schemes is that they do little more than legalise today's noisy technology. In the UK according to one report not one lorry went above the permitted level in GLC tests of hundreds of lorries in 1966 [23]. In 1970 there were apparently only two prosecutions under the 'exceeding maximum permitted sound level' provision of the Motor Vehicles (Construction and Use) Regulations of 1969. Ironically, as if this were not enough of a concession to the motor manufacturers, the standard for lorries was relaxed temporarily in December 1970 from 89 dB(A) to 92 dB(A). Considering that a rise of 3 dB(A) signifies a doubling of sound intensity, Britons can expect little change in lorry noise in the near future, and might even suffer an increase. Standards in other countries reveal the same problem of leniency.

The control of aircraft noise in the United States demonstrates both the positive and negative features of the approach. Under an amendment to the Federal Aviation Act [24], the responsibility for setting aircraft standards was given to the Federal Aviation Administration. The FAA licences aircraft by a process called type certification. This means that before an aircraft type or model can be cleared for flight, it must undergo testing to see if it complies with the various FAA regulations. The new amendment to the law gave the FAA the authority to set noise standards for aircraft it certifies. The agency has already set levels for all new aircraft.

Whether the limits will achieve significant reductions in aircraft noise is questionable for two reasons. First, like their terrestrial counterparts, the aircraft limits are not really very stringent. Second, because they apply only to new types and not to existing fleets, it will be some time before their impact is felt. In a recent report to the President and Congress, the US Environmental Protection Agency makes the startling revelation:

'Projections by the Air Transport Association estimate that by 1975 only 18·6 per cent of the fleet will have been certificated under Part 36, and even this is probably optimistic given present economic conditions that will retard aircraft replacements. Thus, to the extent that it depends upon type certification as presently structured, the noise problem will have been only slightly relieved by 1975 and, indeed, could still be significant as late as 1990' [25].

Such a conclusion is a fine testimony to the victories of the aircraft and airline lobbies.

Regardless of its apparent ineffectiveness, the concept of Part 36 of the FAA rules is novel, for it uses the routine licensing power of the Federal government to achieve environmental goals not necessarily related to the original mission of the agency. While such government power has been used before to advance other social goals such as civil rights, it is rarely found in environmental laws. There is vast untapped potential for all government departments and agencies which engage in contracting, licensing and procurement with the private sector to compel compliance with standards. Governments should exploit this concept.

Asking departments with other missions to enforce special standards can be a double-edged sword, as FAA experience indicates. The FAA has often been criticised for its cosiness with the aircraft manufacturers and the airlines who would not like to pay the full social costs of their activities. Where an agency unsympathetic to amenity questions is also empowered to set standards, and the power is exclusive as it is with the FAA, lenient standards are only to be expected. Ideally an agency more directly concerned with environmental problems, say the Environmental Protection Agency, should set the standards, and the FAA should carry them out in the certification proceedings. More will be said about this later.

Proposals

For the reasons discussed most of the laws have achieved little in the way of actual abatement. I will now outline some proposals which I feel would improve the chances of success of any legal scheme for the abatement of noise. These are of two types: those shifting the approach of control away from remedial towards preventive considerations; and those that strengthen enforcement procedures.

Preventive Law

As we have seen, a major shortcoming of most remedial legal sanctions is that little or no use is made of control technology at the design stage of machinery and equipment. As a consequence, there is little encouragement or advancement for abatement technology. Yet when machinery and consumer goods wear out and are replaced with newly manufactured products as fast as they are today, tremendous potential exists for applying abatement technology at the design and manufacture points. One way to exploit this potential is to create a comprehensive system of emission standards for new products and machinery.

At the very minimum, standards should limit noise and vibration emissions to levels of no proven danger to the health and welfare of humans. Ideally, the standards should include a margin of safety to protect people from levels strongly suspected to be psychologically or physiologically damaging but not yet conclusively proven to be so. Because standards will usually be set by administrative bodies rather than by a court, it will be easier to plan for margins of safety. For it is the nature of the judicial process in most countries to respond (at least theoretically) to what is certain and proven, not to what is speculative, even where research later bears out the speculation. The absence of margins of safety in most existing noise laws reflects a failure to take advantage of the planning potential of administrative regulation and a capitulation to the conservative judicial approach that until a hazard has been proven it cannot be limited or controlled. In an age when the knowledge of the total effects of potential hazards is only gathered long after their release into the environment, this approach to regulation unfairly places an enormous burden upon the public, who must risk exposure to unknown hazards that later prove to be dangerous. Instead of placing the burden of establishing the danger and harm on the potential victims of an environmental hazard, the regulatory scheme should shift the burden to the originator of the hazard, compelling him to prove its innocence before permitting him to release it.

To protect the public and to encourage quieter technologies, standards should increase in severity with time. Lawmakers should not baulk at setting future standards higher than those which present technology is capable of achieving. Only in this way will the existing research energy of large manufacturers be redirected to solving noise emission problems. Traffic noise would markedly decline if manufacturers spent as much time and money on curbing external vehicle noise for the public benefit as on silencing the car's interior. Graduated noise emission standards could assist this.

The only limit on increasingly severe standards should be the estimate of what future technology is capable of achieving. High standards can be set as long as the existence of the technology and the feasibility of its application can safely be predicted at the time of promulgation. There could be subsequent interim hearings on such anticipatory standards evaluating the progress of industry in approaching the new standards. If it were found that the standard makers had predicted wrongly and that, in fact, efforts to comply would be impossible, there could be a provision for an administrative review and modification of these standards. Care would have to be used in designing such an exemptive procedure, lest the value of the graduated emission standard as a key incentive to innovation were lost. Some relevant safeguards are discussed below.

Anticipatory standards have appeared in at least one regulatory context—the United States Clean Air Act of 1970 [26]. The automobile companies were required to remove certain percentages of air pollution emissions by a certain date. When the standard was set, there was no commercially feasible method of complying with the law while producing a saleable product. It was expected on the other hand that with their vast research and development facilities the automakers could attain the standard in the time specified and a year's grace period was allowed.

Emission standards should include a requirement that the abatement feature of any complying product operate over the entire useful life of that product. Accompanied by the enforcement procedures described below, such provision could do much to ensure that quietness was being built into technology rather than simply tacked on to it. The danger of emission standards being negated by this tack-on technology and shoddy manufacturing is not hypothetical. The first automobile air pollution emission standards laid down under Federal law applied only to vehicles at first sale. After the car was sold, the obligation of the manufacturer for the air pollution ceased. Tests of low-mileage automobiles in the US demonstrated the law's failure to achieve long-term durable vehicle emission control. One series of Federal tests revealed that 53 per cent of the 1969 model cars exceeded the limits for that year after 11,000 miles. For some makes the rate soared to 80 per cent.

Standards for new products will begin to tackle the problem of future noise but will do nothing to abate present noise. As we have seen, the regulations setting emission standards promulgated by the Federal Aviation Administration in the US specifically exempt existing aircraft and will only apply to about 18 per cent of the fleet in the US as late as 1975 given present replacement rates. To overcome this sort of problem, a standards scheme should contain provisions applicable to existing noise sources within a fixed time. Such a limitation would speed the transition to new, quieter machines and induce the owners or operators of existing products to add quietening devices.

Along with emission standards should go operating standards. Although jurisdictions already have limits on the operational noise of machinery and equipment written into anti-noise ordinances, these should be expanded and made more specific, as in California. These standards will assure that those owning or operating noise-producing products will use them and maintain them such that their quiet features are preserved. The two types might be combined into a single standard, as the Environmental Protection Agency in the US proposes. Then anyone making, selling or operating the non-complying product would be in violation of the law. Under such a combined approach, care must be taken not to confuse priorities. Emphasis on controlling the noise maker should not lead to neglect of the manufacturers of the noisy product. The first priority must always be to design quiet operation into products before they are sold. In noise control programmes having both sorts of standards—emission and operating—enforcement provisions will play a key role.

Almost as important as the standards is the question of how and by whom they are set. Specific technical limits in a broad regulatory scheme should not be hammered out in the legislative process, but should be decided by an administrative body having flexibility and expertise. Thus the pending legislation in the United States delegates the power of setting noise emission standards to the Environmental Protection Agency and proposals in Britain would authorise the Department of the Environment to make the limits. However, in surrendering standard-making to an administrative department, the legislature must vouchsafe the public interest by vigorous regulation. It has been well documented that government administrators will tend to favour the economic interests they are supposed to regulate unless they are given strong incentives to do otherwise. Because of their isolation from the public—they are neither elected themselves nor appointed by elected officials, and their mouths are often closed by rules of secrecy—government decision-makers tend to be responsive and accountable to those interests with time and money enough to sponsor experts and lobbyists whose job is to persuade the government of their employers' point of view. Trade and industry groups who can afford the expenses involved and who can profit from relaxed standards may lessen the impact of government regulation on their business activities. Where noise is to be regulated the aircraft and motor vehicle manufacturers and transport organisations, as well as the producers of consumer products and industrial equipment, would quietly mount campaigns to protect their industries from what they might consider unreasonable regulation. Of course, these industrial representatives are not evil or sinister. In many cases they believe that what they propose and support in the way of regulation is in the public interest. In some cases it may be. Despite good intentions, and the occasional mutuality of interest with the public, the primary instincts of these organisations are survival, production and profits. When government decisions are based almost exclusively on information, advice and expertise supplied by the business community, violations of the public interest are bound to occur.

The remedy to such lop-sided government decision-making is not simple, though some immediate steps can be taken and have been taken. One is to ensure that the control programme is placed in an agency or department which is capable of administering it. Lawmakers must avoid sending environmental programmes to agencies which have been 'captured' by the very economic interests which the law charges them with regulating. As we have seen, giving the Federal Aviation Administration in the US the power to set aircraft noise standards is one example of this. There are many others. Before the creation of the Environmental Protection Agency, the Federal Water Pollution control programme was administered by a division of the Department of the Interior, an organisation also actively engaged in promoting industrial and mining interests. Likewise, responsibility for the pesticides control laws lay with the Department of Agriculture, an agency long famous for its protection of farmers and the manufacturers of agricultural products. Both these programmes, and the public, suffered incalculably from these allocations. The emergence of government departments devoted, at least on paper, solely to environmental protection is a hopeful sign, but without the additional safeguards discussed below, reorganisation will merely bear out the observation—'*Plus ça change, plus c'est la même chose*'.

Another step that can be taken to correct the imbalance in government decision-making is to place public consumer and environmental representatives on regulatory and advisory boards and councils. Examples of this in Britain are: an environmental journalist on the Noise Advisory Council's Working Party on the Noise Abatement Act; a member of the Consumers' Association on the BSI Technical Sub-committee on Lead in Petrol; and non-industry scientists on the Royal Commission on Environmental Pollution.

Yet another idea is the opening up of the procedure for making standards. In the United States most proposals for the sub-stantive rules and regulations of government departments must be given a public hearing. The Nixon Administration's proposal for a federal noise law provides for public hearings on proposed emission standards, but the Environmental Protection Agency is required to hold them only if requested by a manufacturer whose product is covered by the proposed standard. If anyone else requests a hearing the Administrator of EPA may call it within his discretion. Such provision only exacerbates the existing imbalance in industry and public access to government decision-making. The law should require the agency to hold hearings no matter who requests them. Furthermore, it should require extensive disclosure by an industry which disputes a standard of facts and figures to substantiate its argument. Only in this way can experts—in and out of government—not representing the industry, evaluate the claims.

Generally, environmental laws which give administrators dis-cretionary powers lead to a weakening of the regulatory effort. An official who is not compelled to hold a hearing or promulgate a rule or standard is much more vulnerable to special interest influence. Thus, noise laws should avoid grants of discretionary power as much as possible. An example of what not to do is the provision in the currently proposed noise law which enables but does not require the Environmental Protection Agency to set certain emission standards.

Local v. *Central Controls*

One problem which every government will have to face, if it has not done so already, is how to apportion noise control as between local and central authorities. With regard to noise emission, the problem usually reduces to the question of whether local governments are permitted limits more stringent than those set centrally.

The US debate on this issue is becoming more intense. Impatient with a laggard Federal government, many states are passing comprehensive noise control schemes complete with standards. Trade and industry groups meanwhile are pushing strongly for Federal legislation which will prevent the states from setting emission levels lower than those set at the Federal level. The bill contains such a limitation, known in America as Federal pre-emption, but alternate proposals, backed by state and municipal lobby groups, expressly reject it.

Whether pre-emption should be a part of a central noise law or not is difficult to decide. If pre-emption is chosen, important local prerogatives will be usurped. Since noise tends to be a localised phenomenon, a strong argument can be made for letting local and state governments protect their constituents as they see fit. A uniform standard cannot possibly serve the diverse needs of every noise-weary community. It can be argued that the proliferation of state and local standards will not necessarily delay technological advances. Faced with competing standards, the central government might be moved to pass stricter limits. And industry, in anticipation of conflicting requirements, might design to the highest standard possible.

Opponents of pre-emption point out that the concept has been rejected in other regulatory schemes with few undesirable results. In the US, for instance, the state of California required new automobiles sold within its borders to meet certain air pollution emission limits before Federal law required limits for all cars, while British manufacturers have been equipping cars for export to America with safety and pollution control devices required by US law though not yet mandatory in the United Kingdom. There has been little evidence of a decline in the automobile industry because of the burden of meeting disparate standards.

On the other hand, there is definitely a point at which great diversity of standards becomes counter-productive. Probably one reason why the car industry, despite its grumbling, can comply with different safety requirements in different countries and still remain viable is that the size of the market and the relative cheapness of the modifications justify the additional expenditures. If the market were much smaller, or if several hundred different standards within the same market had to be met, a manufacturer would be disinclined to invest in the research and development necessary to tailor his product to the market's various legal requirements. Where a manufacturer knows that there is one noise standard that he must meet, then compliance becomes easier for him, and he more readily makes the proper investment in order to comply. If the aim of noise law is to encourage investment of this type, and to do it effectively, then uniform, centrally set standards will probably provide surer incentives for quiet technology than a flurry of disparate measures passed locally. However, this will be true only if the size of the market to which the disparate standards apply is small.

Where international markets are involved, the case for pre-emption weakens; countries cannot be expected to yield their sovereignty to an international organisation unless it can be certain that its people will be as closely protected by the international standards as by national standards.

Even when pre-emption occurs at the national level, lawmakers must take great care to make sure that the public interest in vigorous control is not betrayed. Pre-emption and uniformity often mean a regulatory programme which is particularly vulnerable to special interests. It is easier for a manufacturer to try to persuade one central governmental authority than a dozen. Noise laws containing pre-emption language should therefore provide for greater public access to the central decision-making process. Furthermore, pre-emption should not apply to the law's enforcement. Local and other governmental bodies should have full freedom to enforce central standards. Finally the law should permit local governmental units to pass their own standards controlling the use, operation, or movement of products emitting noise. This prerogative not only ensures that special, local problems requiring more stringent control than possible through centrally set emission standards can be dealt with locally, it also assuages the bitterness that pre-emption is bound to cause.

Further Enforcement Proposals

New laws and standards mean nothing unless they are obeyed, and compliance does not occur automatically. In the US, at least, the same lobbies which fight against environmental regulation as it is debated and drafted do not vanish when it becomes law. They merely change gear and focus their efforts on those who must administer the new law, tirelessly working to soften its impact on those they are paid to represent.

Of course, one always hopes for a substantial degree of voluntary compliance with new laws. Enforcement actions are expensive and unpleasant, although probably less so than the pollution they are aimed at abating. Experience shows however that voluntary compliance only occurs when those who must comply find that obeying the law is more economical than ignoring it. Government can encourage this determination by making clear that the law will be vigorously enforced and that violators will be prosecuted. To do this, government officials must have the proper tools. Until effective enforcement measures are put into laws, neither voluntary nor compelled compliance will be achieved, and the laws will be for naught.

Penalties

Penalties which truly deter are essential items in an enforcement arsenal. Existing laws provide for a variety of penalties, including criminal and civil fines, injunctive relief, loss of contracting privileges, and sometimes even imprisonment. Many authorities agree, however, that even when they are applied, fines are usually too small to deter lawbreaking and that the other measures have failed because they are so rarely used.

One solution is to increase fines. The Noise Advisory Council's working group on the Noise Abatement Act has recommended a tenfold increase in the maximum fines under the Act (see paragraph 190 of *Neighbourhood Noise*) and pending Federal legislation in the United States sets very high maximum fines. As a further deterrent, minimum fines should be set and laws should specify that each day of continued violation be treated as a separate offence and subject to a separate fine. When a manufacturer or operator knows he may be heavily fined, he will try harder to comply.

Another solution is to personalise liability in the case of a firm. One reason firms can break the law where an individual might not, is that the corporate entity insulates individual decision-makers from personal responsibility. One proposal before the US Congress provides that 'Whenever a corporation knowingly or wilfully violates ... [the Act], any individual director, officer, or agent of such corporation who knowingly or wilfully authorised, ordered or performed any of the acts or practices constituting in whole or in part such violation shall be subject to such penalties, in addition to the corporation' [27].

If noise emission standards include a requirement that products and machinery be designed and built for continued compliance with the limits, then the penalty structure should support this. The law should extend the manufacturer's liability beyond the point of first sale and make him, in effect, the guarantor of the product's quietness as long as it is properly used and maintained. An operator innocently violating standards by using a defective product and consequently fined should be permitted to claim from the manufacturer. This right would depend on the defendant operator's timely motion in court and his innocence. To prevail, the manufacturer would have the burden of proving that the product met the required standards when sold and would still comply were it not for the negligent or knowing failure of the operator properly to maintain or use the product. If a court found for the operator and ordered indemnification, this would not exempt the manufacturer from separate prosecution for violation of the emission standards.

Placing the burden of proving compliance on the manufacturer would be fair, for in many cases his resources would be greater than the operator's. Subjecting him to double liability would give added incentive to design and build durability into the quietening features of the product.

Extending the manufacturer's liability can also be accomplished administratively by requiring the recall of non-complying products in certain circumstances, in much the same way that automobile makers now recall cars to correct safety defects.

Public Participation in Enforcement

Although they help, heavier penalties do not guarantee compliance with environmental laws. Government officials can still choose not to use the enforcement tools at their disposal, hoping, often vainly, for voluntary compliance. Such inaction does great damage to the law's integrity and to the public interest. For the polluters will learn that they can continue to pollute merely by promising to do better; and the government learns that it can continue dreaming of voluntarism and pretend to enforce the law simply by praising the polluter's good intentions and accepting his promises.

The victims of this game are the general public and traditionally the law has prevented them from breaking it up. Unless new noise laws encourage public initiative and thereby create enforcement incentives for laggard government officials, the game will continue and noise will increase.

A first step is to amend or re-interpret those provisions of existing laws which stand in the way of effective citizen action and invite continued manufacture and use of noisy products. In Britain, the Civil Aviation Act must be changed to give back to the public their common law rights to sue in nuisance against noise from airports. Although several bills to this effect have been introduced in Parliament in recent years none has succeeded in reaching the Statute Book. In addition, the Noise Abatement Act should be amended (i) to remove exemption now enjoyed by some of the country's biggest noisemakers—statutory undertakings—and (ii) to remove the oppressive requirement that three people must complain before a magistrate can act under the public complaint procedures. The working party on the Noise Abatement Act of the Noise Advisory Council has recommended the first change. The second they have left, although the present requirement is an arbitrary and unreasonable burden on the citizen. If a complaint from a single member of the public is unwarranted, then let the magistrate decide this in the course of the proceedings. The public doesn't vote by threes; it shouldn't have to complain by threes.

In all noise law, phrases like 'best practicable means' and 'economically reasonable' should be either deleted or re-interpreted to take into account the full external, social costs of noise. These phrases have too often permitted industrialists to pass to an unwilling public costs which properly should be figured among those of doing business. Of course, industrialists will say that removing these escape phrases from regulatory law will result in standards so strict that factories will close, prices will soar and inflation increase. Although statements like these frighten the public and soften governmental resolve to clean the environment, businessmen may make them with impunity under present law; a wall of secrecy prevents any objective evaluation of industry's true ability to absorb the costs of control in such cases.

Where threats of economic dislocation due to environmental control are real, then they are not to be feared or fought against, for such dislocation tends ultimately towards a more efficient allocation of resources. In its 1970 report to the President, the United States Council on Environmental Quality described the problem as follows:

'When full production costs are included in the prices of final products, the market allocates resources efficiently. If, however, all costs are not included—for example the costs to society of environmental degradation—then the resulting prices of the products are too low. When products are underpriced, consumption of them is higher than it would be if all costs were included. Consequently, compared with other products, too many resources are devoted to their production. To the extent that the costs of preventing undesirable environmental impacts are not reflected in the price of goods and services efficiently, too much waste is produced' [28].

In the United States, the Administration's proposed noise law unfortunately includes the language 'economically reasonable' in defining the standards which the government is authorised to make. Other noise law proposals, however, reject this language. The absence of this sort of phrase will not mean that economics will not enter the picture in setting standards. In the inevitable compromises of government regulation, the cost of control will obviously be considered by officials when serious dislocation becomes a reality. Requiring decisions to be based on traditional economic criteria diverts attention from what should be the chief thrust of environmental laws—the protection of health and welfare—and makes this subsidiary to company finance.

One reason the public has been placed on the sidelines in officialdom's skirmish against pollution is that the public has lacked 'standing'. This meant that one's person or property had to be directly affected by the pollution. For example, 'standing' is a prerequisite in the Noise Abatement Act of 1960 to the bringing of any complaint by members of the public. The Act says the complainants must be occupiers of land or premises and must be aggrieved by a noise nuisance. The public had to rely almost exclusively on its law enforcement officials to see that justice was done.

This ancient obstacle to citizen-action is now being challenged. In the United States, an idea known as the citizen suit has emerged and has already been incorporated into some environmental laws. The citizen suit simply provides that an ordinary member of the public can sue a polluter for violating the law or sue the government for not carrying out its statutory duties. Mere citizenship gives one standing. There is no requirement that the plaintiff be personally affected by the violation or inaction he is suing to correct. In short, the provision makes every member of the public a potential law enforcement official. It promises more than anything else to curb the favouritism towards industrial polluters consistently evinced by government enforcement officials. As a consequence, it will encourage swift compliance with the laws.

Citizen suit provisions are contained in proposals for a new Federal water pollution control law and in those for a Federal noise control law. In both cases the language is borrowed directly from the citizen suit provision which became law in the United States Clean Air Act of 1970 [29]. This provides that any person may commence a civil action in a US District Court against any other person (including a corporate person) who is alleged to be in violation of any emission standard or limitation under the Act, or against the Administrator of the Environmental Protection Agency where he is alleged to have failed to perform any act or duty under the Act which is not discretionary with him. The provision requires that the plaintiff give sixty days' notice to the defendant in either case before commencing the action, except where the alleged violation creates a hazardous situation. It also bars any action against a violator if the government is already prosecuting to require compliance, but the citizen is given the right to intervene in such a case. At the end of the proceedings the court, in its discretion, may award the costs of litigation, including reasonable attorney and expert witness fees to any party.

Though the bare skeleton of the citizen suit remains intact, several of these provisions cut deeply into its effectiveness. The sixty-day delay period is an unfair restraint on the citizen litigant; after all, a policeman is not required politely to inform a shoplifter that in two months' time he will be arrested. More serious is the restriction on suing when the government has commenced an action. The government might settle its case for a weak judgement allowing, for example, an unreasonably long time for the defendant industry to comply with the particular standards or orders it had violated. In such a case, the public would be precluded from bringing a citizen suit.

While better than nothing, the provision giving the court discretion to award costs to any party hardly encourages citizen suits. Ideally, the plaintiff should automatically collect costs if he wins, as in a private civil action. In other cases, the court should be given discretion in awarding costs. However, the citizen plaintiff should never be required to absorb the other side's costs as well as his own. In the first place, his resources will rarely equal those of a manufacturer or operator. Moreover, even if he loses, assuming he has proceeded in good faith from the start, the plaintiff has been doing the government's work for it and should not be penalised.

A good case can be made for awarding the winning public litigant more than simply his costs. In private damage suits, a prevailing plaintiff will sometimes win monetary damages, but in existing citizen suit provisions, any civil fines imposed after a judgement accrue to the government. To encourage wider use of citizen suits and reward the public for its vigilance, the law should provide for a bounty of a portion of the fine levied, say one half, to go to the successful public litigant.

Ironically, the only environmental law in the United States containing a reward provision is the Refuse Act of 1899 (33 USC 407). This law prohibits dumping of refuse (including effluent discharges) into navigable waters of the US without a permit and provides that any member of the public supplying information to the government which leads to the apprehension and conviction of a violator is entitled to one half of any fine imposed. No modern environmental legislation has been so progressive, nor so direct as this simple law of seventy years ago.

Actions against government officials under the citizen suit provisions, of course, depend on the extent of the statutory duties explicitly given to them by the legislation. The old action of *mandamus*, predecessor to this part of the citizen suit and still available in many jurisdictions to compel government officials to discharge their duties, has fallen into disuse because few explicit duties appear in the average regulatory scheme and because judges are likely to interpret ambiguous statutory language as giving discretionary rather than obligatory powers. As pointed out earlier, grants of discretionary power in environmental laws greatly increase the opportunities of special interest lobbies to erode the proper administration and enforcement of these laws. The fact that discretionary duties also immunise government officials from the citizen suit is one more reason to avoid them. Enforcement of the law against violators is too important to be left to the discretion of a government official.

Control laws should also permit members of the public to sue other government departments as well as firms contracting with the government for violations of contract provisions calling for compliance with noise standards. The potential for using government contracting and licensing powers to abate noise has already been discussed. Unless citizens and local and state government officials are permitted to monitor compliance and push directly for enforcement of these contract and licence provisions, little abatement will actually occur. Traditionally, the principle of privity in contract law has prevented interference by outsiders with the contracting parties. But in the matter of contracts with government for services and products, this concept has too often been

used to insulate government and industry from public account-
ability and should be abandoned. A public or government contract
should be subject to public scrutiny and enforcement.

Information and Disclosure

An essential prerequisite for effective public participation is
accurate information. Environmental laws which require broad
public disclosure—by government and by industry—of pollution
facts and figures will lead to an informed public, an enforcement-
minded government and compliance-minded polluters.

Too often secrecy on the part of industry and government has
thwarted citizen attempts to understand environmental problems
and to take action to solve them. For instance, in most countries,
data on industrial water and air pollution emissions have long
been treated as trade secrets and withheld from the public. This is
so even though the public must breathe the air and use the water,
and even though those firms whose figures have become public by
policy, law or court action have not suddenly fallen prey to competition (in the US now, the Clean Air Act requires the public
disclosure of air pollution emissions data). The purpose of this
secrecy is not to protect trade secrets but to protect polluters from
the force of the law and inactive government officials from embarrassment. For without knowing what is going into the air and
water, and who is putting it there, the public is powerless to do
anything substantive about environmental problems. Furthermore, this sort of secrecy does little to inspire the confidence and
trust of the public in government and industry, whose public
relations releases simultaneously exclaim their candour, openness
and concern about environmental problems.

As a matter of high priority, noise laws should require measures
for the disclosure at every stage in the regulatory process of
data about noise characteristics of products and processes subject
to control. At a minimum, manufacturers should be required to
indicate by label on their products how much noise each produces
and be subject to a fine and an injunction for selling non-complying products.

Some jurisdictions have already initiated label requirements.
In the Soviet Union, a regulation (GOST 11970–66 entitled
'Machines: noise characteristics and their measurement'), confirmed in 1966, is aimed at helping Soviet branch industries meet
norms on noise through correct design and lay-out of industrial
plants. The regulation requires emission characteristics of most
new Soviet machinery to be measured while they are in the proto-

type and testing stage. It then requires labels to be put on the machines when they are produced and sent to the plant where they will be used.

Of course labelling should cover more than just industrial machinery. It should also apply to construction and transportation equipment and consumer products. Assuming there is a market for quieter equipment and appliances, the requirement of accurate emission labels should assist both buyers and sellers in that market and create healthy incentives for quieter technology.

Disclosure provisions in noise control laws can and should go beyond labelling. Existing sources, such as industrial works, should be required to publish the noise levels emitted at their boundaries. In applying for planning permission for future or expanded works, a firm should make public those characteristics of its new facility which will have or tend to have an impact on the surrounding environment, including community noise levels. When such vital information becomes a matter of public record, industry and government will be disinclined to take a cavalier attitude towards noise control.

Information which a government collects about a company's environmental record, including noise data, in the course of their monitoring and enforcement duties should also be made public. This is especially true for data collected in the industrial working environment. In most countries, including the US and Britain, the working-man now has *little* right to see reports of government-sponsored studies and measurements of noise and other industrial hazards. The rationale for withholding this information from the men to whom it matters most is a mystery.

Finally, the law should require that a firm which threatens to close a works for environmental reasons, including alleged inability to comply with noise control orders or standards, must make public all the relevant financial and operating data which give rise to the threat. Too often firms will be tempted to use the environment as a convenient pretext for closing down marginal or unprofitable works which would have to be closed in any case. Blaming the environment for the shutdown seriously undermines government clean-up efforts and alienates the worker from the cause of cleaner, quieter surroundings. If a firm knows it will have to open its books to the public, it will not make such threats carelessly. When environmental standards are in fact entirely responsible for the threatened shut-down, it will be in the public interest to have the whole case out in the open. As the American

Jurist Louis Brandeis once quipped, 'Sunlight is the best of disinfectants'.

Noise can be controlled if government, industry and the public determine that it should be. The public is beginning to raise its voice. Government and industry have yet to stir from their unhearing slumber. The solutions offered here to the problem of noise control depend largely on greater attentiveness to the victims of noise pollution—the general public—on the part of government and industry than has been shown in the past. This will come about only when those with the power and the duty to protect the public from excessive noise begin to trust them and enlist their support and understanding in the struggle to curb this environmental hazard. For until government trusts the people, the people cannot be expected to trust the government.

References

1. *Noise—Sound without Value*, Report of the Committee on Environmental Quality of the Federal Council for Science and Technology (September 1968), p. 47.
2. See the following English cases: *Walter* v. *Selfe*, 64 Eng. Rep. 849 (Ch 1851); *Betts* v. *Penge UDC* (1942) 2 KB 154; *Rushmer* v. *Polsue & Alfieri Ltd* (1906) 1 Ch. 234.
3. See English cases: *Polsue & Alfieri Ltd* v. *Rushmer* (1907) AC 121, 122 (opinion of Lord Loreburn); *Halsey* v. *Esso Petroleum Co., Ltd* 1 WLR 683 (QB 1961). This last case is most encouraging, for it shows the potential of the common law to protect the rights of an individual against encroachment by a large multinational company.
4. See the following American cases: *American Airlines, Inc.* v. *City of Audubon Park, Ky.* 407 F 2d 1306 (6th Cir. 1969); *American Airlines Inc.* v. *Town of Hempstead* 398 F 2d 369 (2d Cir. 1968) (cert. denied), 393 US 1017 (1969); *Allegheny Air Lines* v. *Village of Cedarhurst* 238 F 2d 812 (2d Cir. 1956).
5. See Mecklin, 'It's time to turn down all that noise', *Fortune* (October 1969).
6. Report of Committee on Environmental Quality of the Federal Council for Science and Technology, 1968 (see note 1), p. 32.
7. See *The Guardian* for 7/12/71, article entitled 'Damages for noise at work'.
8. 41 USC, s. 35 (1942).
9. 29 USCA, s. 651 *et seq.* (1971).

10. Federal Register, Vol. 36, No. 105 (Saturday, 29 May 1971), p. 10518.
11. M. Goldman, 'Environmental disruption in the Soviet Union', in T. R. Detweiler (ed.), *Man's Impact on the Environment* (McGraw-Hill, 1971).
12. Federal Council for Science and Technology report (see note 1), pp. 34–5.
13. Ibid., loc. cit.
14. R. A. Baron, *The Tyranny of Noise* (New York, 1970), p. 149.
15. US Environmental Protection Agency, *Report to the President and Congress on Noise* (31 December 1971), pp. 4–47.
16. *Neighbourhood Noise*, The Noise Advisory Council, Report by the working group on the Noise Abatement Act (1971), p. 21, para. 98.
17. 42 USC, s. 4331 (Supp. V, 1970).
18. 23 USC, s. 101 (1965).
19. British Standards Institution, British Standard Code of Practice CP3, CH. III: Sound Insulation and Noise Reduction (1960). For a good analysis of noise standards in building codes generally, see 'Statute: a model ordinance to control noise through building code performance standards', *Harvard Journal on Legislation*, Vol. 9 (November 1971).
20. FHA No. 2600, *Minimum Property Standards for Multifamily Housing*, s. M405 (February 1971).
21. California Vehicle Code, s. 27160 (West Supp. 1969).
22. The Administration's bill HR 5275, 92nd Cong. 1st Session, was introduced on 1 March 1971. Several other bills, including HR 6002 and HR 6986, were introduced shortly thereafter. Hearings were held in the House on these and all identical bills before the Subcommittee on Public Health and Environment of the Committee on Interstate and Foreign Commerce on 16, 17, 22, 23, and 24 June 1971. The hearing transcripts and the bills together with all written comments to the Subcommittee can be obtained from the Committee on Interstate and Foreign Commerce under the title of *Noise Control—1971*, Serial No. 92-30.
23. See 'Keep the peace' by Jane Alexander in *New Society* (2 September 1971), p. 404.
24. 88 Stat. 395 (1968).
25. US Environmental Protection Agency, *Report to the President and the Congress on Noise* (31 December 1971), pp. 4–36.
26. 42 USC 1857 *et seq.* and Public Law 91–604, 84 Stat. 1676 (1970 Amendments).
27. See s. 413(d)(2) of HR 6986, 92nd Congress, 1st Session, introduced by Mr Ryan and co-sponsored by twenty other Congressmen. When alternative proposals to that of the Administration are spoken of in this chapter, it is usually this bill which is referred to.
28. *Environmental Quality*, the Second Annual Report of the Council on Environmental Quality (August 1971), p. 102.
29. 42 USC 1857 *et seq.*, s. 304 (1970).

Chapter 12

Conclusions

A. AFTER THE STOCKHOLM CONFERENCE ON THE HUMAN ENVIRONMENT, JUNE 1972

by ALLAN D. McKNIGHT
Lecturer, Civil Service College, London

In 1968, on the initiative of Sweden, the United Nations General Assembly decided to hold a world conference on the human environment, stressing the fact that all living and inanimate things amongst which man dwells are part of a single interdependent system and that man will have no place to turn if he despoils his own surroundings through thoughtless abuse. One hundred and thirteen nations attended the conference which was held in Stockholm in June 1972. All were members of the United Nations or one of its specialised agencies. The Soviet Union and some of the East European countries were notable absentees, refusing to participate on the ground that certain non-members, notably the German Democratic Republic (East Germany) were not being allowed to attend.

Apart from the East European boycott which developed late in the piece, the conference had earlier faced a possible threat of non-co-operation from the developing countries, who regarded a despoiled environment as a natural result of the greed and over-production of the wealthy advanced countries and at one time saw the purpose of the conference as a possible ploy by the advanced countries to inhibit the industrial and agricultural development which the developing countries regard as essential if they are to meet the aspirations of their citizens for a higher standard of living. This apparent conflict was ironed out mostly before the conference assembled. It was generally agreed that there need be no clash between concern for development and concern for the environment and that the developing countries should take full account of environmental factors in formulating their national economic plans so as to avoid the mistakes made in the industrialised countries and thereby enhance the quality of life of their peoples. Development with environmental responsibility is the new watchword; it constantly recurs throughout the conference proceedings; much time was spent in spelling the concept out in definitive terms.

Pollution is only one specific aspect of the human environment. Discussion at the conference ranged broadly and the 109 recommendations covered such fields as (a) the management of human settlements including such problems as family planning, housing, the abolition of squatter areas, human malnutrition and noise emissions, (b) development and the environment, in the course of which the United Nations secretariat was asked to make a study of the extent to which the problems of pollution could be ameliorated by reducing the production of synthetics and substitutes and relying on natural products, (c) the need for general public and specialised technical education, (d) the management of natural resources, particularly the assembly of world genetic banks, both plant and animal, for use by succeeding generations, the preservation of forests and wild life, and the assurance that the development of water, minerals and energy does not produce adverse environmental effects, and, last but not least (e) pollutants.

In approaching this massive agenda the Conference saw four tasks ahead of itself:

1. A declaration which would evidence the political will of all countries henceforth to be concerned with the human environment.
2. An action plan which would contain agreements as to specific action required.
3. The provision of funds.
4. A recommendation as to the institutional machinery necessary to make progress.

The third and fourth will not be examined here; the first and second will be outlined so far as they refer specifically to pollution.

A declaration was adopted although much time was taken up in discussing the prime need to halt nuclear weapon tests. The declaration opens with a philosophical statement dealing with the preservation and enhancement of the human environment in which the following paragraphs appear:

1. Man is both creature and moulder of his environment which gives him physical sustenance and affords him the oppor-

tunity for intellectual, moral, social and spiritual growth. In the long and tortuous evolution of the human race on this planet a stage has been reached when through the rapid acceleration of science and technology, man has acquired the power to transform his environment in countless ways and on an unprecedented scale. Both aspects of man's environment, the natural and the man-made, are essential to his well-being and to the enjoyment of basic human rights—even the right to life itself.

2. The protection and improvement of the human environment is a major issue which affects the well-being of peoples and economic development throughout the world; it is the urgent desire of the peoples of the whole world and the duty of all governments.

3. Man has constantly to sum up experience and go on discovering, inventing, creating and advancing. In our time man's capability to transform his surroundings, if used wisely, can bring to all peoples the benefits of development and the opportunity to enhance the quality of life. Wrongly or heedlessly applied, the same power can do incalculable harm to human beings and the human environment. We see around us growing evidence of man-made harm in many regions of the earth: dangerous levels of pollution in water, air, earth and living beings; major and undesirable disturbances to the ecological balance of the biosphere; destruction and depletion of irreplaceable resources; and gross deficiencies harmful to the physical, mental and social health of man, in the man-made environment; particularly in the living and working environment.

4. A point has been reached in history when we must shape our actions throughout the world with a more prudent care for their environmental consequences. Through ignorance or indifference we can do massive and irreversible harm to the earthly environment on which our life and well-being depend. Conversely, through fuller knowledge and wiser action, we can achieve for ourselves and our posterity a better life in an environment more in keeping with human needs and hopes. There are broad vistas for the enhancement of environmental quality and the creation of a good life. What is needed is an enthusiastic but calm state of mind and intense but orderly work. For the purpose of attaining freedom in the world of nature, man must use knowledge to build in collaboration with nature a better environment. To defend and

improve the human environment for present and future generations has become an imperative goal for mankind—a goal to be pursued together with, and in harmony with, the established and fundamental goals of peace and of world-wide economic and social development.

Then twenty-six principles were elaborated, on which four were specific to pollution and which read as follows:

1. The discharge of toxic substances or of other substances and the release of heat, in such quantities or concentrations as to exceed the capacity of the environment to render them harmless, must be halted in order to ensure that serious or irreversible damage is not inflicted upon ecosystems. The just struggle of the people of all countries against pollution should be supported.
2. States shall take all possible steps to prevent pollution of the seas by substances that are liable to create hazards to human health, to harm living resources and marine life, to damage amenities or to interfere with other legitimate uses of the sea.
3. States have, in accordance with the Charter of the United Nations and the principles of international law, the sovereign right to exploit their own resources pursuant to their own environmental policies, and the responsibility to ensure that activities within their jurisdiction or control do not cause damage to the environment of other states or of areas beyond the limits of national jurisdiction.
4. States shall co-operate to develop further the international law regarding liability and compensation for the victims of pollution and other environmental damage caused by activities within the jurisdiction or control of such states to areas beyond their jurisdiction.

The need for research is stressed and for the free exchange of scientific information.

In pursuance of the principles in the Declaration, committees of the Conference made 109 recommendations to constitute an Action Plan. The UN secretariat describes (Document HE/78/ Rev 1) the recommendations concerning pollutants in these terms:

A variety of international measures to check the rising level of contaminants in air, water, food, the oceans and the human body itself are set out in the recommendations of the Conference on pollutants of international significance.

On the general problem of pollution control, the Conference asked governments to act in concert with one another and with international organisations in planning and carrying out control programmes for pollutants which cross national boundaries. It suggested that the United Nations review and co-ordinate this co-operation and encourage the establishment of mechanisms through which states could consult on the speedy implementation of con-certed abatement programmes.

A number of steps were recommended to gather and assess the information man needs about pollutants in the environment if he is to exercise effective control over them. The Conference proposed an increase in the capability of the United Nations system 'to provide awareness and advance warning of deleterious effects to human health and well-being from man-made pollutants'. The stated aim was to provide such information in a form useful to national policy-makers. The Secretary-General was asked to help governments that wished to use such data in their national planning.

It was recommended that the United Nations work out a pro-cedure for identifying pollutants of international significance and consider appointing expert bodies to assess the exposure, risks, pathways and sources of such pollutants. The Secretary-General was requested to ensure that international programmes were used to monitor the accumulation of hazardous compounds at repre-sentative sites. He was asked to improve the international ac-ceptability of procedures for testing pollutants by developing international test schedules and techniques to permit more meaningful comparisons of data gathered by different nations.

The Conference recommended that United Nations agencies develop agreed procedures for setting safety limits for common air and water contaminants. Governments were asked to take internationally proposed standards into account when estab-lishing national standards for pollutants of international significance.

The Secretary-General was asked to see that research in terres-trial ecology was encouraged, supported and co-ordinated so as to provide knowledge about the behaviour and effects of pollutants on the world's animal and plant life. He was asked to ensure that networks of research centres and biological reserves were desig-nated to facilitate analysis of how ecosystems function under natural and managed conditions.

Governments were asked to make available to one another, through the United Nations, information on their pollution research and control activities. They were also asked to assist other governments to participate in international pollution assessment schemes, and it was suggested that the United Nations examine technical assistance needs in the study of pollution problems.

The Conference also adopted recommendations on specific kinds of pollution. On health effects, it recommended a major effort to develop monitoring and research that would make possible the 'early warning and prevention' of the deleterious effects of pollutants. The WHO was asked to help governments monitor air and water in areas where health might be threatened, to establish environmental health protection standards, and to co-ordinate an international system to correlate medical, environmental and family-history data.

In the related area of food contamination, the FAO was asked to join the WHO in setting up research and monitoring programmes that would provide early information on rising trends of contamination. The Conference suggested that the capabilities of the two organisations to help developing countries in regard to food control be expanded. Increased support was requested for the work of the FAO on international standards for pollutants in food and on a code of ethics for international food trade.

The climatic effects of pollution were dealt with in a recommendation that a network of at least 100 stations be set up to monitor the atmosphere, together with another 10 stations in remote areas to monitor long-term atmospheric trends which might cause meteorological changes. Governments were asked to be mindful of activities that might affect climate, carefully evaluate the likelihood and magnitude of such effects, disseminate their findings in advance, and consult other states when they contemplated or engaged in such activities.

Chemicals in the environment were the subject of two recommendations. Governments were asked to use the 'best practicable means' to minimise the release of toxic or dangerous substances, especially persistent ones such as heavy metals (including mercury) and organochlorine compounds (including those found in DDT and other insecticides) 'until it has been demonstrated that their release will not give rise to unacceptable risks or unless their use is essential to human health or food production, in which case appropriate control measure should be applied'. The Secretary-General was asked to develop plans for an International Registry of Data on Chemicals in the Environment, based on a collection of production figures on the most harmful chemicals and data about the environmental behaviour of the most important man-made chemicals from factory to ultimate disposal or recirculation.

As to contamination from radio-activity, the Conference recommended that governments explore the feasibility of developing a registry of release of radio-active materials. It also suggested expanded co-operation on problems of radio-active wastes.

As most of the world's oceans belong to no single nation but are the common heritage of mankind, the Conference devoted much attention to marine pollution. Governments were asked to act quickly to control all significant sources of marine pollution, including the land-based sources which supply most of the pollutants found in the ocean. The Secretary-General was requested to provide guidelines which governments might wish to take into account when developing such measures. Governments were asked to give collective endorsement to a set of principles on the control of marine pollution as guiding concepts for two international conferences scheduled for next year on the law of the sea and on marine pollution.

Articles for a draft convention on ocean dumping, developed during preparatory meetings in London, Ottawa and Reykjavik, are the subject of another recommendation. The Conference asked that the draft be sent for information and comments to the United Nations Committee on the Peaceful Uses of the Sea-Bed and the Ocean Floor beyond the Limits of National Jurisdiction, and for adoption at a conference to be convened by the United Kingdom before November. Governments were asked to ensure control over ocean dumping by their nationals anywhere or by any person in areas under their jurisdiction. They were requested to accept and implement available instruments on control over maritime sources of marine pollution, and to ensure that those instruments were complied with by ships flying their flags or in areas under their jurisdiction.

Governments were asked to participate in efforts to control all maritime sources of marine pollution, particularly with the aim of eliminating all deliberate pollution by oil from ships by the mid-1970s. Radio-active pollution from nuclear surface ships and submarines was singled out for special attention in this connection. Also governments were asked to recognise that, 'in some circumstances, the discharge of residual heat from nuclear and other power stations may constitute a potential hazard to marine ecosystems'.

The Conference urged support for several specific programmes to assess marine pollution: the Global Investigation of Pollution in the Marine Environment, the Integrated Global Ocean Station System, the gathering of statistics on potential marine pollutants, the work of the Intergovernmental Oceanographic Commission (IOC), an annual review of harmful chemical substances in the oceans, the preparation of guidelines for tests to evaluate the toxicity of pollutants, studies on the effects of marine pollutants on man and other organisms, and a study of the possibility of establishing an international institute for tropical marine studies. The intergovernmental bodies concerned were asked to promote the monitoring of marine pollution.

The IOC was requested to consider the strengthening of information exchange activities concerning marine pollution, and to initiate a scientific information referral capability in this area. The Secretary-General was requested to seek more funds for training and other assistance to help developing countries participate in international marine research, monitoring and pollution control programmes. The Conference suggested that the United Nations environmental machinery ensure that advice on marine pollution problems was provided to governments.

Too much should not be expected too quickly for the United Nations machinery grinds extremely slow. But at least a start has been made in what the Swedish delegate designated as a dialogue between peoples and nations. The Secretary-General referred to political decisions having been taken which will enable the community of nations to act together in a manner consistent with the earth's physical interdependence. It is necessary to wait and see how much can be accomplished on a world scale.

Some broad conclusions may be drawn as to what action is required in law and administration relating to pollution.

Firstly, Britain has no reason to be complacent. Although her legislation is claimed to be in advance of most other countries, it is still incomplete, quixotic, showing all the signs of hasty reaction to extreme situations, and far from certain to be observed. In particular, such legislative standards as impose on the producers of pollution an obligation to take 'the best practical steps' to minimise that pollution, allow for subjective interpretation, which usually means 'the cheapest possible steps' which amount to little more than doing nothing. Standards need to be laid down after due assessment of scientific and technical capability to control pollution, the disadvantage flowing from the release of the pollutants, and the economic cost of taking remedial action. All these steps are a necessary preliminary to either national or international action. If determined nationally, they can, based as they are on scientific investigation, be urged by Britain as the basis for international standards. If the latter, when ultimately adopted, fix standards different from British national standards, the latter may be easily amended to conform. In any event, standards need to be reviewed in the light of experience, new knowledge and technical development.

Secondly, either before or at the time the standards are legislatively adopted, power must be taken to review the design of all new plants before construction commences in order to establish that the design will meet the standards applicable. There must be government inspection during construction to ensure the plant is built to the design which has been approved. At the same time programmes must be drawn up for the modification of all existing plant to enable them to meet the new standards within a specified period.

Thirdly, the agencies responsible for ensuring compliance with the new legislative standards must be adequately staffed. The prevailing philosophy that law enforcement is to be obtained in 'cosy club' discussions with potential polluters must change and there must be a greater exercise of strong pressure by the inspectors and a greater willingness to prosecute.

Fourthly, enforcement agencies must be equipped with sophisticated instruments which automatically monitor plant discharges. These must be tamper proof. In this way the number of inspectors will be kept to a minimum, law enforcement will be non-intrusive, and the chance detection of non-compliance will be high. But at the same time even the most sophisticated instruments need strong, effective and independent human support.

Fifthly, penalties for breaches of the legislative standards must be substantial. Trumpery pecuniary penalties will not be sufficient. Parliament must also consider whether the responsible officers and managers of artificial corporations should not be made liable to imprisonment. Parliament must also state its intention that breaches of the new legislative standards confer a right of action for civil damages on each and every person suffering damage to person or property by such a breach, for the fear of a multiplicity of civil actions will act as a powerful pressure in favour of observance of the standards.

such measures requires both the political will and the administrative expertise to put them into effective use.

A scientific problem in the area of air pollution often discussed is the so-called greenhouse effect. Caused by fossil fuel combustion, the CO_2 concentration of the air rises and may cause a trapping of the sun's heat to raise the overall temperature of the atmosphere. The thermodynamic calculation at this stage of the development of knowledge is not sufficiently accurate enough for policy making, though the problem is certainly one of scientific interest. The majority of scientific questions relating to pollution lie in this fairly long-term scale. The substitution of fusion or solar energy for fossil fuel combustion are further examples.

The fundamental problem for science then, as it is related to pollution (or in the wider sense technological disamenity), is that of developing administrative and social control methods for regulating technological advance and economic growth without stultifying scientific invention [4]. Closely linked to this is the determination of socially desirable goals for R and D [5].

It is self-evident that the resources of the world are ultimately finite and that an increasing population and increasing consumption by that population will result in crisis situations unless mechanisms are fostered for using man's inventiveness for finding more efficient substitute resources, methods and materials. Hence the widespread interest and debate on the limits to growth [6, 7]. Value free science is slowly but surely being forced to adjust to the value laden organisation of the industrial and industrialising world. The main scientific aspect of pollution problems lies squarely in this area. This is the problem of anticipating and thereby avoiding the worst deleterious impacts of science and technology at the earliest stage possible. The proposed methods are conveniently summarised under the title of Technology Assessment [8, 9]. This is an attempt to set up at various organisational levels of both government and industry mechanisms of review, analysis and control which will anticipate the deleterious outcomes of any innovation while preserving its benefits, economically and socially. Principles 17 and 18 of the Stockholm Conference read, respectively, as follows: 'Appropriate national institutions must be entrusted with the task of planning, managing or controlling the environmental resources of states with a view to enhancing environmental quality,' and 'Science and technology, as part of their contribution to economic and social development must be applied to the identification, avoidance and control of environmental risks and the solution of environmental problems and for the common good of mankind.' Thus are closely linked the traditional arbitrary role of the legal system and the rational scheme of science. As Lady Jackson put it in her lecture at Stockholm, 'At last, in this age of ultimate scientific discovery, our facts and our morals have come together to tell us how we must live.' The discussions of the foregoing chapters show that such communion is difficult though desirable. Much is made [10] of the major incompatibilities of our systems of production and our environmental system. On the other hand it can be argued fairly forcibly [11] that historical analogies show man to be often capable of surmounting apparent catastrophic situations and improving both his material living standard and his quality of life.

The combination of hopes and fears implicit in the whole debate on the human environment must act as a spur to greater effort towards the integration of scientific thought into political action.

References

1. J. J. P. Staudinger, *Disposal of Plastics Waste and Litter* (Chem. Soc., London, 1970).
2. S. L. Kwee and J. S. R. Mullender, *Growing Against Ourselves* (Lexington Books, 1972).
3. Council on Env. Quality, *The Economic Impact of Pollution Control* (Environmental Protection Agency, Washington, March 1972).
4. C. Freeman, 'Technology assessment in its social setting', *Stud. General,* Vol. 24 (1971), pp. 1038–50.
5. C. Freeman *et al.,* 'The goals of R & D in the 1970s', *Science Studies,* Vol. 1, No. 3 (1971), pp. 357–406.
6. D. Meadows *et al., Limits to Growth* (Potomac Associates, Washington, 1971).
7. Sussex Group, *Malthus on a Computer* (Chatto & Windus for the Sussex Univ. Press, 1973).
8. C. Sinclair, 'Technology assessment in the UK', *Technology Assessment,* Vol. 1, No. 2 (1972).
9. National Academy of Engineering, *A Study of Technology Assessment* (USGPO 1969).
10. B. Commoner, *The Closing Circle* (Cape, 1972).
11. Sussex Group, op. cit., Chapter 3.

Index

Unless otherwise stated legislative acts are British